Proceedings of the

Gökova Geometry-Topology Conference 2010

Editors

Selman Akbulut
Department of Mathematics, Michigan State University
East Lansing, MI 48824, USA

Denis Auroux
Department of Mathematics, University of California
Berkeley, CA 94720, USA

Turgut Önder
Department of Mathematics, Middle East Technical University
06531 Ankara, Turkey

GGT Gökova Geometry-Topology Conferences

International Press
www.intlpress.com

Published in 2011 by International Press, Somerville, Massachusetts, U.S.A. under license from Gökova Geometry-Topology Conferences.

ISBN 978-1-57146-226-8

Printed in the United States of America.

17th GÖKOVA GEOMETRY-TOPOLOGY CONFERENCE
May 31 – June 4, 2010

Participants of the 17th Gökova Geometry-Topology Conference

PREFACE

This volume contains articles from the seventeenth Gökova Conference which took place during the last week of May 2010. We thank all the authors for these nice papers. A special thanks goes to F. Bourgeois for delivering the minicourse on contact homology, we also thank participants for making this a very lively conference. This volume also contains an interesting article surveying the solution of the celebrated Arf-Kervaire invariant problem by Hill, Hopkins and Ravenel. We thank TUBITAK and NSF for funding this conference, and we thank IP for printing and distributing these proceedings. We also thank Selahi Durusoy for putting these proceedings together, and thank Hotel Yücelen (which is located on the scenic shores of the Gökova Bay) for supporting and hosting this conference.

April 2011 The Editors

CONTENTS

Proceedings of 17th Gökova
Geometry-Topology Conference
pp. 1 – 20

Lectures on Symplectic and Contact Homology

Frédéric Bourgeois

Abstract. We shall give the geometric definition of several holomorphic curve invariants for symplectic and contact manifolds, such as contact homology, symplectic homology and some of their variants. We shall then explain the relationship between these theories, leading to a common algebraic framework. These relations can also be used to translate structural results from one of these invariants to the other ones. This can be illustrated with the effect of some geometric operations on these invariants, such as Legendrian surgery.

Contents

1. Symplectic and contact homology

1.1. Symplectic and contact manifolds

Let X be a smooth manifold of dimension $2n$. A 2-form ω on X is called **symplectic** if it is closed (i.e., $d\omega = 0$) and nondegenerate (i.e., $\omega^{\wedge n} \neq 0$ everywhere). The pair (X, ω) is called a **symplectic manifold**.

Let Y be a smooth manifold of dimension $2n-1$. A hyperplane distribution ξ on Y is called a **contact structure** if, for any 1-form α such that $\xi = \ker \alpha$ on some open subset $U \subset Y$, we have $\alpha \wedge d\alpha^{\wedge(n-1)} \neq 0$ on U. The pair (Y, ξ) is called a **contact manifold**. A 1-form α such that $\xi = \ker \alpha$ on Y is called a **contact form**.

To any contact form α on Y, one can associate a vector field R_α called the **Reeb vector field** and characterized by

$$\begin{cases} \imath(R_\alpha)d\alpha &= 0, \\ \alpha(R_\alpha) &= 1. \end{cases}$$

This vector field strongly depends on the choice of a contact form for a given contact structure.

Even though the above defintions are apparently very different, contact manifolds can be thought of as the odd dimensional counterpart of symplectic manifolds. Observe indeed that, if α is a contact form on (Y, ξ), then $d\alpha$ restricts to a symplectic form on ξ. This symplectic form depends on the choice of α only up to a nonvanishing factor. We now describe two constructions illustrating the interplay between contact and symplectic geometry.

Given a contact manifold (Y, ξ), we can construct a symplectic manifold called the **symplectization** of (Y, ξ). It is the manifold $\mathbb{R} \times Y$, equipped with the symplectic form $d(e^t\alpha)$, where t is the coordinate on the $\mathbb{R}$ factor and α is a contact form for ξ. This symplectic manifold does not depend on the choice of α.

On the other hand, let Y be a hypersurface in a symplectic manifold (X, ω). A vector field v on X is called **Liouville** if $\mathcal{L}_v\omega = \omega$. The hypersurface Y is said to be **of contact type** if there exists a Liouville vector field v on X defined near Y and transverse to Y. In that case, the 1-form $\alpha = \imath(v)\omega$ restricts to a contact form on Y. In particular, if $Y = \partial X$ is of contact type with a transverse Liouville vector field v pointing outside X, we say that (X, ω) has a **convex** boundary. If v is pointing inwards, we say that (X, ω) has a **concave** boundary. If a symplectic manifold (X, ω) has a boundary $\partial X = \partial_+ X \cup \partial_- X$ such that $Y_+ = \partial_+ X$ is convex and $Y_- = \partial_- X$ is concave, we say that (X, ω) is a **symplectic cobordism** from Y_+ to Y_-.

If (X, ω) has a convex boundary, we can complete X to a manifold $\widehat{X}$ given by $X \cup_{\partial X \simeq \{0\} \times Y} \mathbb{R}^+ \times Y$ equipped with the symplectic form $\widehat{\omega}$ given by ω on X and by $d(e^t\alpha)$ on $\mathbb{R}^+ \times Y$.

1.2. Complex structures

Many symplectic and contact invariants can be constructed using a compatible (almost) complex structure. In particular, these allow to define pseudo-holomorphic curves, which were introduced in symplectic geometry by Gromov [14].

On a smooth manifold X of dimension $2n$, an **almost complex structure** is an endomorphism $J : TX \to TX$ such that $J^2 = -I$. An almost complex structure J on a symplectic manifold (X, ω) is said to be **compatible** with ω if

$$\omega(Jv, Jw) = \omega(v, w) \qquad \text{and} \qquad \omega(v, Jv) > 0$$

for any $v, w \in TX$ with $v \neq 0$.

It is a well-known fact that the set of compatible almost complex structures on a symplectic manifold (X, ω) is nonempty and contractible. Therefore, the tangent bundle

TX can be considered as a complex vector bundle up to homotopy. In particular, we can define its **first Chern class** $c_1(X, \omega) \in H^2(X, \mathbb{Z})$.

We say that an almost complex structure J on X is **integrable** if (X, J) is a complex manifold, or more explicitly if X admits an atlas with complex coordinates in which J is represented by i and with biholomorphic coordinate changes.

On a contact manifold (Y, ξ), a **complex structure** J is an endomorphism $J : \xi \to \xi$ such that $J^2 = -I$. Such a complex structure J is said to be **compatible** with $d\alpha$ if

$$d\alpha(Jv, Jw) = d\alpha(v, w) \qquad \text{and} \qquad d\alpha(v, Jv) > 0$$

for any $v, w \in \xi$ with $v \neq 0$.

Similarly, the set of compatible almost complex structures on a contact manifold (Y, ξ) is nonempty and contractible, so that the first Chern class $c_1(\xi) \in H^2(Y, \mathbb{Z})$ can be defined.

A compatible complex structure J on (Y, ξ) can be extended to a compatible almost complex structure on the symplectization $(\mathbb{R} \times Y, d(e^t\alpha))$ by $J\frac{\partial}{\partial t} = R_\alpha$. Like the Reeb vector field, this extension strongly depends on the choice of a contact form α for ξ.

Example 1.1. We now introduce our main class of examples for these lectures. A **Stein manifold** is a complex manifold (X, J) admitting a proper, complex embedding in $\mathbb{C}^N$ for some $N > 0$.

A smooth function $\phi : X \to \mathbb{R}$ is called **exhausting** if it is proper and bounded from below. It is called **strictly plurisubharmonic** if the 2-form $-dd^{\mathbb{C}}\phi$ is symplectic, where $d^{\mathbb{C}}\phi = d\phi \circ J$. These notions allow to formulate an intrinsic characterization of Stein manifolds due to Grauert [13]: a Stein manifold is a complex manifold (X, J) which admits an exhausting, strictly plurisubharmonic function ϕ. In particular, Stein manifolds are symplectic and admit a Liouville vector field which is transverse to the regular level sets of ϕ.

Note that since the above conditions on ϕ are open, we can assume that it is Morse. It then follows from the above definition that the critical points of ϕ have index at most $n = \frac{1}{2}\dim X$. Eliashberg [10] used this to give a topological characterization of Stein manifolds: an open smooth manifold X of dimension $2n > 4$ with an almost complex structure J and an exhausting Morse function ϕ with critical points of index at most n admits a Stein structure. More precisely, J is homotopic through almost complex structures to an integrable complex structure J' such that ϕ is strictly plurisubharmonic.

A Stein manifold (X, J, ϕ) is called **subcritical** if the critical points of ϕ have index strictly less than $n = \frac{1}{2}\dim X$.

1.3. Floer homology

Let (X, ω) be a closed symplectic manifold of dimension $2n$. In this context, a smooth function $H : S^1 \times X \to \mathbb{R}$ will be called a **(time-dependent) Hamiltonian function**. For any $\theta \in S^1$, we write $H_\theta = H|_{\{\theta\}\times X}$.

To any (time-dependent) Hamiltonian function H, one can associate a (time-dependent) vector field X^θ_H, called **Hamiltonian vector field** and characterized by

$$\imath(X^\theta_H)\omega = dH_\theta.$$

A closed curve $\gamma : S^1 = \mathbb{R}/\mathbb{Z} \to X$ is called a **1-periodic orbit** of X^θ_H if

$$\dot\gamma(\theta) = X^\theta_H(\gamma(\theta)),$$

for all $\theta \in S^1$. In local coordinates, the above differential equation coincides with Hamilton's equations for classical mechanics.

We say that a 1-period orbit γ of X^θ_H is **nondegenerate** if the graph of the time 1 flow $\varphi^1_H : X \to X$ of X^θ_H in $X \times X$ is transverse to the diagonal $\Delta \subset X \times X$ at point $(\gamma(0), \gamma(0))$. In that case, after choosing a symplectic trivialization of TX along γ, one associates to γ an integer $\mu_{CZ}(\gamma)$ called the **Conley-Zehnder index** of γ. See [19] for a definition of this index.

The **Novikov ring** Λ_ω of (X, ω) over $\mathbb{Z}$ is a completion of the group ring $\mathbb{Z}[H_2(X,\mathbb{Z})]$; it consists of elements

$$\sum_{A \in H_2(X,\mathbb{Z})} \lambda_A e^A$$

such that the number of nonzero coefficients λ_A with $\omega(A) < C$ is finite for any $C \in \mathbb{R}$. The ring Λ_ω is graded by $|e^A| = -2\langle c_1(X,\omega), A\rangle$.

Let $H : S^1 \times X \to \mathbb{R}$ be a generic Hamiltonian, so that all 1-periodic orbits of X^θ_H are nondegenerate. For each such contractible orbit γ, we choose a spanning disk D_γ and use it to induce a symplectic trivialization of TX along γ. We define a module $FH(X, \omega)$ freely generated over Λ_ω by the contractible 1-periodic orbits of X^θ_H. This module is graded by the Conley-Zehnder index : $|\gamma| = \mu_{CZ}(\gamma)$.

This module will be equipped with a differential in order to imitate the Morse complex for the **action functional**

$$\mathcal{A}_H : C^\infty_{\mathrm{contr}}(S^1, X) \to \mathbb{R} : \gamma \mapsto -\int_{D^2} \sigma^*\omega - \int_{S^1} H(\theta, \gamma(\theta))\, d\theta,$$

where $\sigma : D^2 \to X$ is a smooth homotopy between $\sigma_{\partial D^2} = \gamma$ and a constant loop in X. This action functional is well-defined if (X, ω) is symplectically aspherical, i.e., $\langle [\omega], \pi_2(X)\rangle = 0$. The critical points of $\mathcal{A}_H$ are exactly the contractible 1-periodic orbits of X^θ_H.

To define gradient trajectories in Morse theory, we need to choose a Riemannian metric on the domain of the Morse function. In this case, such a metric will be induced by a (time dependent) compatible almost complex structure J_θ, $\theta \in S^1$, on (X, ω). The equation for gradient trajectories $u : \mathbb{R} \times S^1 \to X$ for $\mathcal{A}_H$ can then be written formally as:

$$\frac{\partial u}{\partial s} + J_\theta(u)\left(\frac{\partial u}{\partial \theta} - X^\theta_H(u)\right) = 0, \tag{1}$$

with asymptotic conditions

$$\lim_{s\to-\infty} u(s,\theta) = \gamma_-(\theta), \qquad \lim_{s\to\infty} u(s,\theta) = \gamma_+(\theta), \tag{2}$$

for some contractible 1-period orbits γ_-, γ_+ of X_H^θ.

These equations form a well-defined elliptic problem, but unlike in Morse theory these do not define a flow. Note that it is not necessary to have a flow in order to define a differential complex. Indeed, the construction of Floer homology is based on the **moduli spaces of Floer trajectories** $\mathcal{M}^A(\gamma_-, \gamma_+)$ consisting of solutions u of (1) and (2), such that the homology class of the sphere obtained by attaching the image of u with the capping disks D_{γ_-} and D_{γ_+} coincides with $A \in H_2(X, \mathbb{Z})$.

If (X,ω) is aspherical, we can find a generic J_θ so that $\mathcal{M}^A(\gamma_-, \gamma_+)$ is a smooth manifold of dimension $|\gamma_-| - |\gamma_+| - |e^A|$. This manifold has an $\mathbb{R}$-action defined by $(\tau \cdot u)(s,\theta) = u(s+\tau, \theta)$. This collection of moduli spaces can be equipped with coherent orientations, i.e. orientations which are compatible with the operation of gluing Floer trajectories. When the quotient $\mathcal{M}^A(\gamma_-, \gamma_+)/\mathbb{R}$ has dimension zero, it consists of finitely many rigid Floer trajectories, each equipped with a sign. We denote by $\#\mathcal{M}^A(\gamma_-, \gamma_+)/\mathbb{R}$ the algebraic count of these trajectories.

We define a **differential** $d_{\mathrm{FH}} : FH_*(X,\omega) \to FH_{*-1}(X,\omega)$ by

$$d_{\mathrm{FH}}\gamma_- = \sum_{\gamma_+, A} \#\mathcal{M}^A(\gamma_-, \gamma_+)/\mathbb{R} e^A \gamma_+.$$

Note that the above sum can be infinite, because $\cup_{\omega(A)\le C}\mathcal{M}^A(\gamma_-, \gamma_+)/\mathbb{R}$ is compact for all $C < \infty$. However, the union of moduli space $\mathcal{M}^A(\gamma_-, \gamma_+)/\mathbb{R}$ over all homology classes $A \in H_2(X,\mathbb{Z})$ may be noncompact. This is why a partial completion is required in the definition of the ring Λ_ω.

Then $d_{\mathrm{FH}} \circ d_{\mathrm{FH}} = 0$ and the homology $F\mathbb{H}(X,\omega) = H(FH(X,\omega), d_{\mathrm{FH}})$ is independent of H and J. It is called the **Floer homology** of (X,ω). Moroever,

$$F\mathbb{H}_*(X,\omega) \simeq H_{*+n}(X, \Lambda_\omega)$$

by comparing the Floer and the Morse complexes for appropriate Hamiltonian functions H.

This theory was originally developed by Floer [12] in order to prove a homological version of the **Arnold conjecture**, stating that the number of contractible 1-periodic orbits for a time-dependent Hamiltonian on a closed symplectic manifold (X,ω) is bounded below by the sum of the Betti numbers of X.

1.4. Symplectic homology

Let (X,ω) be a symplectic manifold with convex boundary $Y = \partial X$. We shall apply the construction of Floer homology on the completed manifold $(\widehat{X}, \widehat{\omega})$. Since this manifold is not compact, it is necessary to specify the behavior of H and J on $\mathbb{R}^+ \times Y$.

We first choose $H : S^1 \times \widehat{X} \to \mathbb{R}$ so that

(i) $H < 0$ is a C^2-small Morse function on X;

(ii) $H(\theta, t, p) = h(t)$ is strictly increasing on $\mathbb{R}^+ \times Y$, such that $h''(t) - h'(t) > 0$ and, for t large enough, $h(t) = ae^t + b$,

with $a > 0$ is generic, so that X_H^θ has no 1-periodic orbit outside a compact set.

We then require that

$$J_\theta \xi = \xi, \qquad J_\theta \frac{\partial}{\partial t} = R_\alpha, \qquad \text{and } J_\theta \text{ is independent of } t,$$

when $t \in \mathbb{R}^+$ is large enough.

In this context, the homology of the Floer complex is denoted by $S\mathbb{H}(H, J)$. This graded module is in fact independent of J, but it does depend on H through the asymptotic slope $a > 0$.

In order to remove this dependence we use a direct limit and define the **symplectic homology** of (X, ω) by

$$S\mathbb{H}(X, \omega) = \lim_{\overrightarrow{H}} S\mathbb{H}(H),$$

where the direct limit uses monotonicity morphisms $\sigma : S\mathbb{H}(H_1) \to S\mathbb{H}(H_2)$ for $H_1 < H_2$, counting solutions of the Floer equation with an s-dependent Hamiltonian interpolating between H_1 and H_2.

To understand how $S\mathbb{H}(H)$ depends on the slope $a > 0$, consider the 1-periodic orbits of X_H^θ. These fall in two categories:

(i) the orbits in X are critical points of H;
(ii) all orbits contained in $\mathbb{R}^+ \times Y$ correspond to closed Reeb orbits on Y, because $X_H^\theta = -e^{-t} h'(t) R_\alpha$. We therefore obtain all closed Reeb orbits with period in $[0, a]$.

These two types of generators for the complex $SH(X, \omega)$ of symplectic homology are distinguished by the values of the action functional $\mathcal{A}_H$: there exists $\varepsilon > 0$ such that

(i) if $\mathcal{A}_H(\gamma) < \varepsilon$, then γ is a critical point of H in X. These generators therefore form a subcomplex $SH^-(X, \omega) \subset SH(X, \omega)$.
(ii) if $\mathcal{A}_H(\gamma) > \varepsilon$, then γ corresponds to a closed Reeb orbit on Y. These orbits generate the quotient complex $SH^+(X, \omega) = SH(X, \omega)/SH^-(X, \omega)$.

It turns out that the complex $(SH^-(X, \omega), \partial)$ coincides with the Morse complex of H on X, so that

$$S\mathbb{H}_*^-(X, \omega) \simeq H_{*+n}(X, \partial X; \Lambda_\omega).$$

Therefore, the short exact sequences of complexes induces in homology the exact triangle

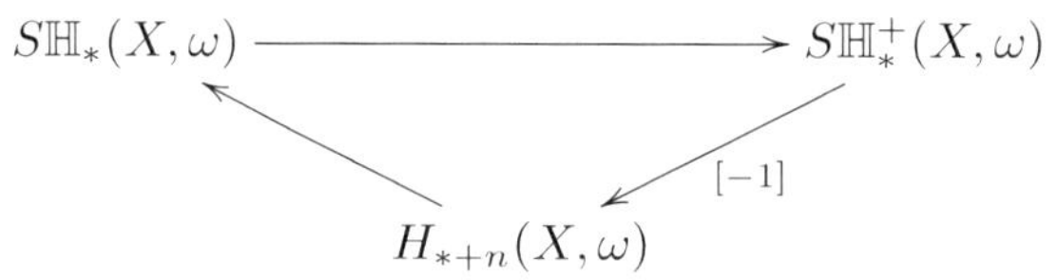

For more details about the definition and the properties of symplectic homology, see [18].

Example 1.2. Let ω_0 be the standard symplectic structure on the ball $X = B^{2n}$. Then

$$\mathbb{SH}(B^{2n}, \omega_0) = 0.$$

To see this, if r denotes the radial coordinate, we have $e^t = r^2$ on $\widehat{X} = \mathbb{R}^{2n}$, so that we can choose the Hamiltonian H of the form $Ar^2 - B$ on $\mathbb{R}^{2n}$, with very large $A, B > 0$. For generic $A > 0$, we obtain a single 1-periodic orbit γ which is constant at the origin $r = 0$. Since this generator satisfies $|\gamma| \to \infty$ as $r \to \infty$, the direct limit over H (i.e., as $A \to \infty$) gives a vanishing symplectic homology in each degree.

In fact, it was proved by Cieliebak [7] that symplectic homology remains zero for any subcritical Stein manifold (X, ω). Inserting this result in the exact triangle, we obtain $\mathbb{SH}^+_*(X, \omega) \simeq H_{*+n-1}(X, \partial X; \Lambda_\omega)$.

1.5. Contact homology

Contact homology is a holomorphic curves invariant for a contact manifold (Y, ξ). It is a small part of a much more general framework for holomorphic curves invariants of contact manifolds and of symplectic manifolds with convex and concave boundaries. This framework is called Symplectic Field Theory and was introduced by Eliashberg, Givental and Hofer [11]. In fact, the specific invariant we are going to consider in these lectures is often called **linearized contact homology**, but since we are not going to consider other variants of contact homology in these notes, we will simply write contact homology.

Let us make a tentative definition of it based on the above constructions, noting that the complex SH^+ is generated by closed Reeb orbits in Y. Ignoring the symplectic filling (X, ω), let us concentrate on the symplectization $(\mathbb{R} \times Y, d(e^t\alpha))$ and set $H = 0$.

Tentative definition. We define a complex generated by closed Reeb orbits (of any period). The differential counts holomorphic cylinders, i.e. maps $F = (a, f)$ from $\mathbb{R} \times S^1$ to $\mathbb{R} \times Y$ satisfying $df \circ j = J \circ dF$, such that

$$\lim_{s \to \pm\infty} a(s, \theta) = \pm\infty, \qquad \lim_{s \to \infty} f(s, \theta) = \gamma_+(T_+\theta), \qquad \lim_{s \to -\infty} f(s, \theta) = \gamma_-(T_-\theta),$$

for some parametrizations γ_+ and γ_- of closed Reeb orbits with periods T_+ and T_-. The collection of such holomorphic cylinders, modulo biholomorphisms of the domain, is called **moduli space of holomorphic cylinders**.

This tentative definition has the following two problems :

(i) We need to exclude "bad orbits" in order to guarantee that the moduli spaces of holomorphic curves are orientable [3]. For any closed Reeb orbit γ, we denote by γ_1 the underlying simple orbit, i.e., the orbit with the same image and the minimal period. We say that the orbit γ is **good** if $\mu_{CZ}(\gamma) - \mu_{CZ}(\gamma_1)$ is even, otherwise we say that γ is **bad**.

(ii) The moduli spaces of holomorphic cylinders are compactified by adding configurations of holomorphic curves that are more general than cylinders. For instance, a 1-parameter family of holomorphic cylinders may degenerate in a rigid pair-of-pants with two negative punctures, attached to a rigid holomorphic plane and a vertical cylinder. In order to take these more general degenerations into account, we have to replace the above moduli spaces of holomorphic cylinders with **moduli spaces of capped holomorphic curves**. These moduli spaces $\mathcal{M}_c^A(\gamma_+, \gamma_-)$ consist of tuples of maps $(F_+, F_1, \dots, F_k)$, modulo biholomorphisms of their domains. The map

$$F_+ = (a_+, f_+) : \mathbb{R} \times S^1 \setminus \{y_1, \dots, y_k\} \to \mathbb{R} \times Y$$

satisfies $dF_+ \circ j = J \circ dF_+$ and

$$\lim_{s \to \pm\infty} a_+(s, \theta) = \pm\infty, \qquad \lim_{z \to y_i} a(z) = -\infty,$$

$$\lim_{s \to \infty} f_+(s, \theta) = \gamma_+(T_+\theta), \qquad \lim_{s \to -\infty} f_+(s, \theta) = \gamma_-(T_-\theta),$$

$$\lim_{\rho \to 0} f_+(y_i + \rho e^{2\pi i \vartheta}) = \gamma_i(T_i \vartheta),$$

for some parametrizations $\gamma_+, \gamma_-, \gamma_1, \dots, \gamma_k$ of closed Reeb orbits with periods $T_+, T_-, T_1, \dots, T_k$. For $i = 1, \dots, k$, the maps

$$F_i : \mathbb{C}P^1 \setminus \{\infty\} \to \widehat{X}$$

satisfy $dF_i \circ j = J \circ dF_i$. Moreover, a neighborhood U_∞ of $\infty \in \mathbb{C}P^1$ is mapped by F_i to $\mathbb{R}^+ \times Y$, so that we can write $F_i|_{U_\infty} = (a_i, f_i)$. These components satisfy

$$\lim_{|z| \to \infty} a_i(z) = \infty, \qquad \lim_{\rho \to \infty} f_i(\rho e^{2\pi i \vartheta}) = \gamma_i(T_i \vartheta).$$

For each $i = 1, \dots, k$ we fix a conformal identification of $\mathbb{C}P^1 \setminus \{\infty\} \simeq \mathbb{C}$ with $T_{y_i}\mathbb{R} \times S^1 \simeq \mathbb{C}$ which is compatible with the limits of f_+ and f_i along γ_i. Note that the number of possible conformal identifications is the multiplicity of the orbit γ_i. We finally require that the homology class of the surface obtained by attaching the images of $F_+, F_1, \dots, F_k$ with the capping disks D_{γ_-} and D_{γ_+} coincides with $A \in H_2(X, \mathbb{Z})$.

The above definition of capped holomorphic curves is a particular case of the more general concept of holomorphic building [2], which is very natural in Symplectic Field Theory.

In favorable circumstances, there exists generic J such that $\mathcal{M}_c^A(\gamma_+, \gamma_-)$ is a smooth manifold of dimension $\mu_{CZ}(\gamma_+) - \mu_{CZ}(\gamma_-) + 2\langle c_1(X, \omega), A\rangle$. Note that unlike in symplectic homology, there are no simple assumptions to guarantee the existence of such a generic J. If such J does not exist, then one has to use other types of perturbations, such as in the theory polyfolds developed by Hofer, Wysocki and Zehnder [15, 16, 17].

Taking the above difficulties into account, we can modify the tentative definition of contact homology.

Actual definition. We define the **contact complex** as the module $CH(Y, \alpha)$ freely generated over Λ_ω by all good closed orbits of the Reeb field R_α. The grading of a generator γ is defined by $|\gamma| = \mu_{CZ}(\gamma) + n - 3$.

We define a **differential** $d_{\mathrm{CH}} : CH_*(Y,\alpha) \to CH_{*-1}(Y,\alpha)$ by

$$d_{\mathrm{CH}}\gamma_+ = \sum_{\gamma_-,A} \#\mathcal{M}_c^A(\gamma_+,\gamma_-)/\mathbb{R}\ e^A\gamma_-. \tag{3}$$

The homology $H(CH(Y,\alpha), d_{\mathrm{CH}})$ of this chain complex is independent of J and of the contact form α for ξ. It is denoted by $\mathbb{CH}(X,\omega)$ and called **linearized contact homology** of (Y,ξ), with respect to the filling (X,ω). In some cases, the resulting invariant depends only on the contact manifold (Y,ξ) and not on the choice of a symplectic filling. It is then usual to denote this invariant by $\mathbb{CH}(Y,\xi)$.

Example 1.3. The symplectic manifold $(X,\omega) = (B^{2n},\omega_0)$ has convex boundary $(Y,\xi) = (S^{2n-1},\xi_0 = \ker\alpha_0)$ with $\alpha_0 = \frac{1}{2}\sum_{i=1}^n (x_i dy_i - y_i dx_i)|_{S^{2n-1}}$. The perturbed contact form $\widetilde{\alpha} = \frac{1}{2}\sum_{i=1}^n a_i(x_i dy_i - y_i dx_i)|_{S^{2n-1}}$ with $a_i > 0$ defines a contact structure diffeomorphic to ξ_0. If the coefficients $a_1,\ldots,a_n$ are linearly independent over $\mathbb{Q}$, then the images of the simple closed orbits $\gamma_j, j = 1,\ldots,n$ of the Reeb field $R_{\widetilde{\alpha}}$ are given by

$$\{(x_1,y_1,\ldots,x_n,y_n) \in S^{2n-1} \subset \mathbb{R}^{2n} \mid x_i = y_i = 0, i \neq j\}.$$

The grading of γ_j and their multiples is always even, so that $d_{\mathrm{CH}} = 0$ independently of the symplectic filling. We obtain

$$\mathbb{CH}_k(S^{2n-1},\xi_0) = \begin{cases} \mathbb{Z} & \text{if } k \geq 2n-2 \text{ is even,} \\ 0 & \text{otherwise.} \end{cases}$$

More generally, we can compute contact homology for the convex boundary (Y,ξ) of a subcritical Stein manifold (X,ω).

Theorem 1.1 (M.-L. Yau [21]). *Let (X,ω) be a subcritical Stein manifold such that $c_1(X,\omega) = 0$. Then*

$$\mathbb{CH}_k(X,\omega) = \bigoplus_{m\geq 0} H_{k-2m+2}(X,\partial X;\Lambda_\omega).$$

2. Relation between symplectic and contact homology

2.1. Exact triangles

Both invariants $\mathbb{SH}^+(X,\omega)$ and $\mathbb{CH}(X,\omega)$ are based on solutions of a Cauchy-Riemann type equation, with asymptotes corresponding to closed Reeb orbits in Y. The examples show that these two invariants are not isomorphic. There are indeed a number of differences in their definitions:

(i) The differential for SH counts objects in $\widehat{X}$, while the differential for CH counts objects living mainly in $\mathbb{R}\times Y$.
However, by "stretching the neck" of $\widehat{X}$ along $\{0\}\times Y$, we split off a copy of $\mathbb{R}\times Y$ in which the objects counted by the differential will mainly live in.

(ii) The symplectic complex uses a Hamiltonian function H, while $H = 0$ for the contact complex.

But we can define maps between these complexes, counting solutions of the Floer equation with a Hamiltonian function interpolating between H and 0 in $\mathbb{R} \times Y$.

(iii) Unlike SH, the complex CH is generated by closed orbits with multiplicities. But in fact, the multiples $\gamma^{(1)}, \gamma^{(2)}, \dots$ of a closed Reeb orbit are realized as 1-periodic orbits of X_H^θ, for increasing values of t.

(iv) The bad orbits are excluded from CH, but not from SH. But a bad orbit gives rise to a pair of generators in SH; these cancel in homology, if we use Novikov coefficients over $\mathbb{Q}$.

(v) A 1-periodic orbit of X_H^θ has a fixed starting point, corresponding to $\theta = 0$, while there is no favorite basepoint on a closed Reeb orbit. This leads to an important difference between the complexes SH and CH, since any closed Reeb orbit gives rise to a pair of generators in SH. Intuitively, CH looks like a version of SH modulo an S^1-action.

The relation between $S\mathbb{H}^+(X, \omega)$ and $C\mathbb{H}(X, \omega)$ is decribed by the following result.

Theorem 2.1 (Bourgeois, Oancea [4]). *Let (X, ω) be a symplectically aspherical manifold with convex boundary (Y, ξ) such that there exists a regular J to define contact homology. Then there is an exact triangle with coefficients in $\Lambda_\omega \otimes_{\mathbb{Z}} \mathbb{Q}$:*

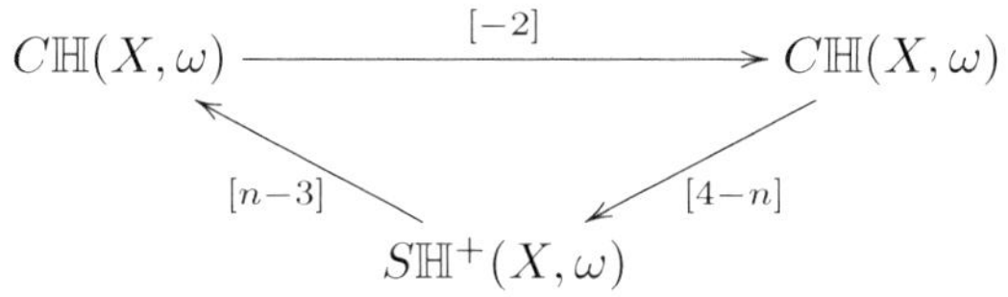

Note that the above exact triangle is analogous the the Gysin exact sequence for a circle bundle $S^1 \hookrightarrow V \to B$:

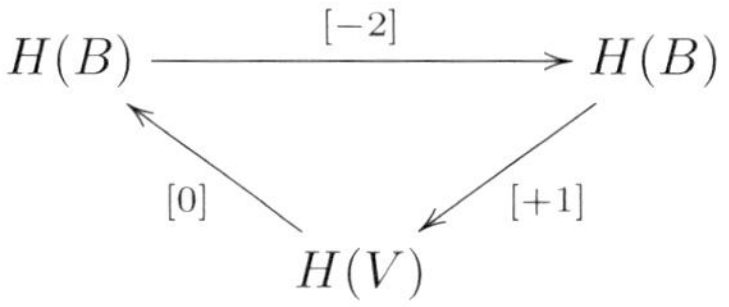

Since $H(B) \simeq H^{S^1}(V)$, this analogy motivates the construction of an S^1-equivariant version of symplectic homology. To that end, let us first recall the Borel construction of S^1-equivariant homology.

Borel construction. Let M be a smooth manifold with an S^1-action. Let ES^1 be a contractible topological space with a free S^1-action. Such a topological space is unique up to homotopy equivalence. We denote by $M \times_{S^1} ES^1$ the quotient of the product $M \times ES^1$ by the diagonal S^1-action. Note that the latter is free, since the action on the second factor is free. The S^1-equivariant homology of M is defined by $H^{S^1}(M) = H(M \times_{S^1} ES^1)$.

It is related to the ordinary homology of M by the exact triangle

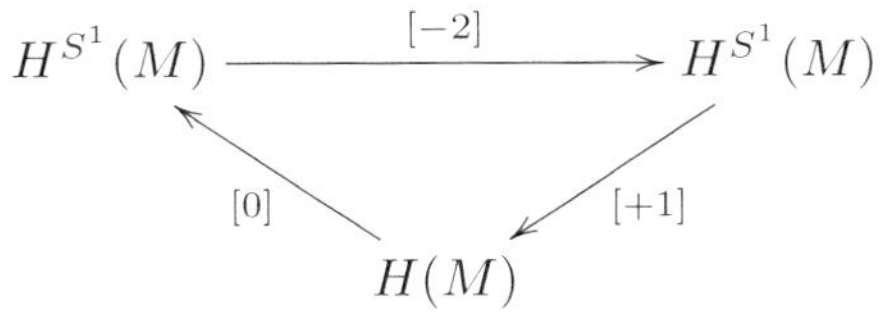

Let us now adapt this classical construction to symplectic homology, following a sketch by Viterbo [20].

Symplectic construction. The sphere S^{2N+1} with the Hopf S^1-action can be viewed as a finite dimensional approximation of ES^1, because we have the S^1-equivariant inclusions $\ldots \hookrightarrow S^{2N+1} \hookrightarrow S^{2N+3} \hookrightarrow \ldots \hookrightarrow S^\infty = ES^1$.

We need the following geometric data:

(i) A family of compatible almost complex structures J^λ_θ, $\theta \in S^1$, $\lambda \in S^{2N+1}$, such that $J^{\tau\cdot\lambda}_{\theta+\tau} = J^\lambda_\theta$ for all $\tau \in S^1$.

(ii) A smooth function $H : S^1 \times \widehat{X} \times S^{2N+1} \to \mathbb{R}$ such that $H(\theta, t, p, \lambda) = ae^t + b(\lambda)$ for t sufficiently large on $\mathbb{R}^+ \times Y$, for some function $b : S^{2N+1} \to \mathbb{R}$, and $H(\theta + \tau, x, \tau \cdot \lambda) = H(\theta, x, \lambda)$ for any $\tau \in S^1$.

We then define a parametrized version of the action functional $\mathcal{A}_H$ from $C^\infty_{\mathrm{contr}}(S^1, \widehat{X}) \times S^{2N+1}$ to $\mathbb{R}$ by

$$\mathcal{A}_H(\gamma, \lambda) = -\int_{D^2} \sigma^*\omega - \int_{S^1} H(\theta, \gamma(\theta), \lambda)\, d\theta.$$

It follows from the assumptions of H and J that $\mathcal{A}_H$ is invariant under the S^1-action defined by $\tau \cdot (\gamma, \lambda) = (\gamma(\cdot + \tau), \tau \cdot \lambda)$.

In analogy with symplectic homology, we define the **S^1-equivariant symplectic complex** $SH^{S^1,N}(X, \omega)$ as the module freely generated over Λ_ω by the circles $\tau \cdot (\gamma, \lambda)$ of critical points of $\mathcal{A}_H$. These consist of a 1-periodic orbit γ of $X^\theta_{H_\lambda}$ and of $\lambda \in S^{2N+1}$ such that $\int_{S^1} \frac{\partial H}{\partial \lambda}(\theta, \gamma(\theta), \lambda)\, d\theta = 0$.

The **differential** $\partial : SH^{S^1,N}_*(X, \omega) \to SH^{S^1,N}_{*-1}(X, \omega)$ counts rigid solutions (modulo the S^1-action) of

$$\left\{ \begin{array}{rcl} \frac{\partial u}{\partial s} + J^\theta_{\lambda(s)}(u)\left(\frac{\partial u}{\partial \theta} - X^\theta_{H_{\lambda(s)}}(u)\right) & = & 0, \\ \dot{\lambda} - \int_{S^1} \frac{\partial H}{\partial \lambda}(\theta, u(s,\theta), \lambda)\, d\theta = 0, & & \end{array} \right.$$

such that

$$\lim_{s \to \pm\infty} (u(s, \theta), \lambda(s)) = (\gamma_\pm(\theta), \lambda_\pm).$$

The homology $H(SH^{S^1,N}(X, \omega), \partial)$ of this complex is independent of J and is denoted by $S\mathbb{H}^{S^1,N}(H)$. In order to obtain an S^1-equivariant symplectic invariant, we use a double direct limit to define the **S^1-equivariant symplectic homology** of (X, ω) by

$$S\mathbb{H}^{S^1}(X, \omega) = \varinjlim_N \varinjlim_H S\mathbb{H}^{S^1,N}(H),$$

where the first direct limit uses morphisms $S\mathbb{H}^{S^1,N}(X,\omega) \to S\mathbb{H}^{S^1,N+1}(X,\omega)$ induced by the inclusions $S^{2N+1} \hookrightarrow S^{2N+3}$.

The construction of this invariant and its relation to the ordinary symplectic homology are provided by the following result.

Theorem 2.2 (Bourgeois, Oancea [5]). *Let (X,ω) be a symplectically aspherical manifold with convex boundary. Then its S^1-equivariant symplectic homology $S\mathbb{H}^{S^1}(X,\omega)$ is well-defined as a symplectic invariant and it fits into the exact triangle with coefficients in $\Lambda_\omega \otimes_{\mathbb{Z}} \mathbb{Q}$:*

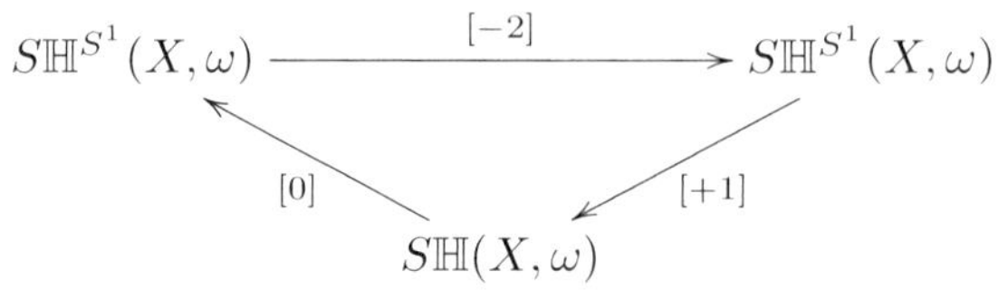

Like in ordinary symplectic homology, one can define a subcomplex $SH^{S^1,N,-}(X,\omega)$ with generators having action less than a small $\epsilon > 0$. This gives rise to an exact triangle involving $S\mathbb{H}^{S^1}(X,\omega)$, $S\mathbb{H}^{S^1,+}(X,\omega)$ and $S\mathbb{H}^{S^1,-}(X,\omega) \simeq H(X,\partial X;\Lambda_\omega) \otimes H(BS^1)$, where $BS^1 = ES^1/BS^1$. Combining with the above theorem, we obtain an exact triangle of exact triangles with coefficients in $\Lambda_\omega \otimes_{\mathbb{Z}} \mathbb{Q}$:

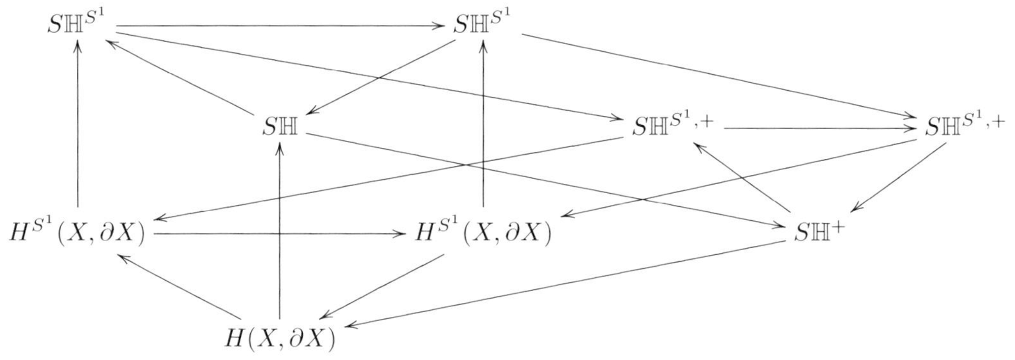

The S^1-equivariant version of symplectic homology and its Gysin triangle relating it to the ordinary symplectic homology can lead to a better understanding of contact homology $C\mathbb{H}(X,\omega)$ and its relation to symplectic homology $S\mathbb{H}^+(X,\omega)$. This is the object of the following result.

Theorem 2.3 (Bourgeois, Oancea [6]). *Let (X,ω) be a symplectically aspherical manifold with convex boundary (Y,ξ), such that there exists a regular J to define contact homology. Then there is an isomorphism of exact triangles with coefficients in $\Lambda_\omega \otimes_{\mathbb{Z}} \mathbb{Q}$:*

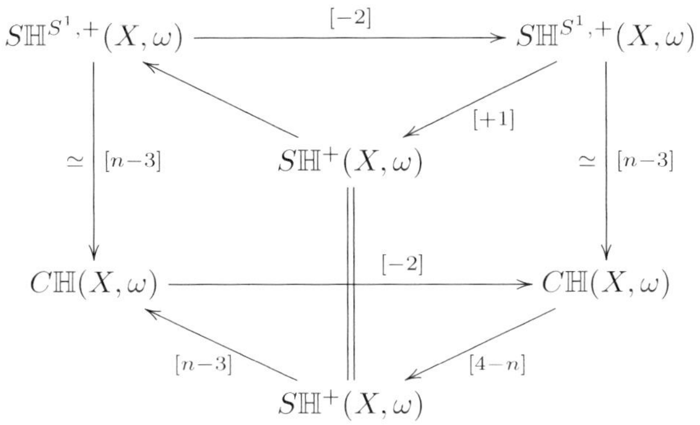

This result shows in particular that contact homology $C\mathbb{H}(X,\omega)$ can always be replaced with $S\mathbb{H}^{S^1,+}(X,\omega)$. The disadvantage is that the latter invariant has a more complicated definition, due to the S^{2N+1} factor from the Borel construction. The advantage is that we can always find a regular J for this latter invariant on (X,ω) symplectically aspherical, unlike for contact homology.

Example 2.1. Let (X,ω) be a subcritical Stein manifold with convex boundary (Y,ξ) such that $c_1(X,\omega)=0$. Since $S\mathbb{H}(X,\omega)=0$, we infer that

$$S\mathbb{H}_k^+(X,\omega) \simeq H_{k+n-1}(X,\partial X;\Lambda_\omega).$$

Similarly, we have $S\mathbb{H}^{S^1}(X,\omega)=0$, so that $S\mathbb{H}_k^{S^1,+}(X,\omega) \simeq H_{k+n-1}^{S^1}(X,\partial X;\Lambda_\omega)$.

Since the S^1-action on X is trivial, this implies that

$$C\mathbb{H}_{k+n-3}(X,\omega) \simeq S\mathbb{H}_k^{S^1,+}(X,\omega) \simeq \big(H(X,\partial X;\Lambda_\omega)\otimes H(BS^1)\big)_{k+n-1}.$$

In other words, we get

$$C\mathbb{H}_k(X,\omega) \simeq \bigoplus_{m\geq 0} H_{k-2m+2}(X,\partial X;\Lambda_\omega)$$

as before.

2.2. Common algebraic formalism

The complexes SH and SH^+ for symplectic homology admit an alternative definition which is formulated in terms of generators for the contact complex CH. This common algebraic formalism gives the possibility of a unified treatment of these three invariants. More details about the complexes described in this section can be found in [1].

Let us start with the alternative definition of SH^+. We define $\widecheck{CH}(X,\omega)$ as the module generated by all (good and bad) closed Reeb orbits in Y. We also define $\widehat{CH}(X,\omega)$ to be $\widecheck{CH}(X,\omega)[1]$, i.e., the same as the above module but with grading shifted by 1. Given a generator γ of $CH(Y,\xi)$, we denote by $\check{\gamma}$ and $\widehat{\gamma}$ the corresponding generators in $\widecheck{CH}(X,\omega)$ and in $\widehat{CH}(X,\omega)$ respectively. The **alternative SH^+ complex** is given by

$$SH^+(X,\omega) = \widecheck{CH}(X,\omega) \oplus \widehat{CH}(X,\omega).$$

We denote by $d_{\widecheck{CH}} : \widecheck{CH}_*(X,\omega) \to \widecheck{CH}_{*-1}(X,\omega)$ the differential defined by equation (3) where the summation is now over all (good and bad) orbits, and similarly for $d_{\widehat{CH}} : \widehat{CH}_*(X,\omega) \to \widehat{CH}_{*-1}(X,\omega)$.

We define a linear map $d_M : \widehat{CH}_*(X,\omega) \to \widecheck{CH}_{*-1}(X,\omega)$ by

$$d_M \widehat{\gamma} = \begin{cases} 0 & \text{if } \gamma \text{ is good,} \\ \pm 2\check{\gamma} & \text{if } \gamma \text{ is bad.} \end{cases}$$

On each closed Reeb orbit γ in Y, we fix a point P_γ. We define moduli spaces $\widetilde{\mathcal{M}'}^A_c(\gamma,\gamma')$ and $\widetilde{\mathcal{M}''}^A_c(\gamma,\gamma')$ as follows. The moduli space $\widetilde{\mathcal{M}'}^A_c(\gamma,\gamma')$ consists of elements $(F_+, F_1, \ldots, F_k)$ in $\mathcal{M}^A_c(\gamma,\gamma')$ for which there exists $\theta_0 \in S^1$ such that

$$\lim_{s\to\infty} f_+(s,\theta_0) = P_\gamma \qquad \text{and} \qquad \lim_{s\to-\infty} f_+(s,\theta_0) = P_{\gamma'}.$$

The moduli space $\widetilde{\mathcal{M}''}^A_c(\gamma,\gamma')$ consists of pairs of elements

$$((F^1_+, F^1_1, \ldots, F^1_{k_1}), (F^2_+, F^2_1, \ldots, F^2_{k_2})) \in \mathcal{M}^{A_1}_c(\gamma,\gamma'')/\mathbb{R} \times \mathcal{M}^{A_2}_c(\gamma'',\gamma')$$

with $A_1 + A_2 = A$ and for some closed Reeb orbit γ'' such that the points

$$\lim_{s\to-\infty} f^1_+(s,\theta^1_0), \qquad \lim_{s\to\infty} f^2_+(s,\theta^2_0) \qquad \text{and} \qquad P_{\gamma''}$$

lie on γ'' in the cyclic order induced by the direction of the Reeb field, where $\theta^1_0, \theta^2_0 \in S^1$ are defined by

$$\lim_{s\to\infty} f^1_+(s,\theta^1_0) = P_\gamma \qquad \text{and} \qquad \lim_{s\to-\infty} f^2_+(s,\theta^2_0) = P_{\gamma'}.$$

Finally, let $\widetilde{\mathcal{M}}^A_c(\gamma,\gamma') = \widetilde{\mathcal{M}'}^A_c(\gamma,\gamma') \cup \widetilde{\mathcal{M}''}^A_c(\gamma,\gamma')$. Under favorable circumstances as for contact homology, and for a generic choice of the points P_γ, this moduli space is a smooth manifold of dimension $\mu_{CZ}(\gamma) - \mu_{CZ}(\gamma') + 2\langle c_1(X,\omega), A\rangle - 1$.

We define a linear map $\delta : \widecheck{CH}_*(X,\omega) \to \widehat{CH}_{*-1}(X,\omega)$ by

$$\delta\check{\gamma} = \sum_{\gamma',A} \#\widetilde{\mathcal{M}}^A_c(\gamma,\gamma')/\mathbb{R}\; e^A \widehat{\gamma}'.$$

The **differential** $d_{SH^+} : SH^+_*(X,\omega) \to SH^+_{*-1}(X,\omega)$ is defined, with respect to the above decomposition of SH^+, by

$$d_{SH^+} = \begin{pmatrix} d_{\widecheck{CH}} & d_M \\ \delta & d_{\widehat{CH}} \end{pmatrix}.$$

With these definitions, the homology $H(SH^+(X,\omega), d_{SH^+})$ is isomorphic to the symplectic homology $\mathbb{SH}^+(X,\omega)$ defined in section 1.4. The generators $\check{\gamma}$ and $\widehat{\gamma}$ of SH^+ correspond to nondegenerate 1-periodic Hamiltonian orbits obtained from the closed Reeb orbit γ by a time-dependent perturbation. The part d_M of the differential is responsible for the elimination of bad orbits over rational coefficients, as explained in section 2.1. The differentials $d_{\widetilde{CH}}$ and $d_{\widehat{CH}}$ restricted to good orbits then coincide with the contact homology differentials. Finally, the δ part of the differential induces the contact homology endomorphism of degree -2 in the exact sequence in Theorem 2.1.

We now give the alternative definition of SH. Let $\tilde{f} : X \to \mathbb{R}$ be a Morse function such that $\tilde{f}|_{\partial X}$ is constant. Let $(\mathrm{Morse}(\tilde{f}), d_{\mathrm{Morse}})$ be the Morse complex of $\tilde{f}$. The **alternative *SH* complex** is given by

$$SH(X,\omega) = SH^+(X,\omega) \oplus \mathrm{Morse}(\tilde{f})[-n].$$

Given a closed Reeb orbit γ and a critical point p of $\tilde{f}$, we define a moduli space $\mathcal{M}(\gamma, p)$ consisting of maps

$$F : \mathbb{C}P^1 \setminus \{\infty\} \to \widehat{X}$$

satisfying $dF \circ j = J \circ dF$ modulo biholomorphisms of the domain preserving 0. Moreover, a neighborhood U_∞ of $\infty \in \mathbb{C}P^1$ is mapped by F to $\mathbb{R}^+ \times Y$, so that we can write $F|_{U_\infty} = (a, f)$. These components satisfy

$$\lim_{|z|\to\infty} a(z) = \infty, \qquad \lim_{\rho\to\infty} f(\rho e^{2\pi i\vartheta}) = \gamma(T\vartheta),$$

for some parametrization of the closed Reeb orbit γ with period T. We also require that $F(0)$ belongs to the unstable manifold of p for the Morse function $\tilde{f}$ with respect to a fixed Riemannian metric on X. The gradient trajectories of $\tilde{f}$ are extended to $\widehat{X}$ using integral curves of $\frac{\partial}{\partial t}$ on $\mathbb{R}^+ \times Y$.

Let $\theta : SH^+(X,\omega) = \widetilde{CH}(X,\omega) \oplus \widehat{CH}(X,\omega) \to \mathrm{Morse}(\tilde{f})$ be the linear map defined by

$$\theta\widehat{\gamma} = 0, \qquad \theta\check{\gamma} = \begin{cases} \sum_{p\in\mathrm{Crit}(\tilde{f})} \#\mathcal{M}(\gamma,p)\, p & \text{if } \gamma \text{ is good,} \\ 0 & \text{if } \gamma \text{ is bad.} \end{cases}$$

The **differential** $d_{SH} : SH_*(X,\omega) \to SH_{*-1}(X,\omega)$ is defined, with respect to the above decomposition of SH, by

$$d_{SH} = \begin{pmatrix} d_{SH^+} & 0 \\ \theta & d_{\mathrm{Morse}} \end{pmatrix}.$$

With these definitions, the homology $H(SH(X,\omega), d_{SH})$ is isomorphic to the symplectic homology $\mathbb{SH}(X,\omega)$ defined in section 1.4.

3. Effect of Legendrian surgery

The common formalism described in the previous section can be used to describe simultaneously the effect of a Legendrian surgery on Y (i.e., attaching a critical handle to X) for the three invariants $\mathbb{CH}$, $\mathbb{SH}$ and $\mathbb{SH}^+$.

The Legendrian surgery operation is a particular case of the more general contact surgery. The latter consists of attaching a handle to X along an isotropic sphere of (Y, ξ). For an explicit model of the handles and the attaching procedure, see [22]. Legendrian surgery corresponds to the critical case, i.e., an attachment along a Legendrian sphere Λ in (Y, ξ). Subcritical Stein manifolds are the result of subcritical handle attachments to the standard symplectic ball (B^{2n}, ω_0). As hinted at the end of section 1.4, symplectic homology $S\mathbb{H}(X, \omega)$ is not affected by subcritical handle attachments. On the other hand, general Stein manifolds can be obtained from subcritical Stein manifolds by Legendrian surgery along a collection of Legendrian spheres. As we will see in this section, the three holomorphic curve invariants that we consider in these notes are deeply affected by such an operation.

For simplicity, we will consider only the case of Legendrian surgery along a single Legendrian sphere. In order to understand why the effect of Legendrian surgery differs so much from the effect of subcritical surgeries, let us compare the Reeb dynamics on the boundary of the corresponding handles:

(i) On the boundary $D^k \times S^{2n-k-1}$, $k < n$ of a subcritical handle, we can find:
  - Closed Reeb orbits foliating the sphere $\{0\} \times S^{2n-2k-1}$. These do not contribute to symplectic homology, for the same reason as on S^{2n-1}, like in example 1.2.
  - Reeb trajectories leaving the handle. After a perturbation of the isotropic sphere to which the handle is attached, the return time of these trajectories to the handle can be made arbitrarily long, by reducing the size of attaching locus of the handle around the isotropic sphere. For this reason, closed Reeb orbits that are only partially contained in the handle do not contribute to $S\mathbb{H}$, $S\mathbb{H}^+$ or $C\mathbb{H}$.

(ii) On the boundary ST^*D^n of a critical handle, the Reeb dynamics coincide with the geodesic flow on the disk D^n. Therefore, we can find:
  - No closed Reeb orbits in this critical handle, since all Reeb trajectories project to a piece of line moving across D^n.
  - Reeb trajectories leaving the handle. These can return to the handle after a bounded time, since generically a Legendrian sphere Λ will admit Reeb chords, i.e., integral curves of the Reeb field starting and ending on Λ. Since geodesic segments can join any pair of points on $\partial D^n \simeq \Lambda$, one can easily get convinced that the closed Reeb orbits that are created by a Legendrian surgery along Λ are in bijective correspondence with cyclic tuples of Reeb chords of Λ.

This last category of closed Reeb orbits should contribute to our holomorphic curve invariants. Moreover, these orbits depend on the interactions of the Reeb dynamics with the Legendrian sphere Λ. It is therefore to be expected that the effect of Legendrian surgery will be described in terms of a holomorphic curve invariant of Λ, defined in the same spirit as symplectic and contact homology.

We define $LH(\Lambda)$ as the noncommutative unital algebra freely generated over $\Lambda_\omega \otimes_{\mathbb{Z}} \mathbb{Q}$ by the Reeb chords of Λ. This algebra can be equipped with a differential d_{LH} mapping $LH_*(\Lambda)$ to $LH_{*-1}(\Lambda)$ satisfying the Leibniz rule and counting some holomorphic disks in $\mathbb{R} \times Y$ with boundary on $\mathbb{R} \times \Lambda$. The corresponding homology $L\mathbb{H}(\Lambda) = H(LH(\Lambda), d_{LH})$ is an invariant of the Legendrian isotopy class of Λ. For a definition of this invariant for Legendrian submanifolds, see [8, 9].

Cyclic version. We now define a variant of the complex $(LH(\Lambda), d_{LH})$, which plays an important role to describe the effect of Legendrian surgery on $C\mathbb{H}(X, \omega)$. We denote by $LH^+(\Lambda) = LH(\Lambda)/\langle 1 \rangle$ the module generated by words of Reeb chords for Λ. Let $P : LH^+(\Lambda) \to LH^+(\Lambda)$ be the linear map defined by

$$P(c_1 \dots c_k) = (-1)^{|c_k|(|c_1| + \dots + |c_{k-1}|)} c_k c_1 \dots c_{k-1}.$$

We define $LH^{\mathrm{cyc}}(\Lambda) = LH^+(\Lambda)/\mathrm{im}(I - P)$. It is the module generated by cyclic words of Reeb chords for Λ. Note that, since $\langle 1 \rangle$ and $\mathrm{im}(I - P)$ are subcomplexes, the differential d_{LH} induces a differential d_{cyc} on $LH^{\mathrm{cyc}}(\Lambda)$. The corresponding homology $L\mathbb{H}^{\mathrm{cyc}}(\Lambda) = H(LH^{\mathrm{cyc}}(\Lambda), d_{\mathrm{cyc}})$ is again an invariant of the Legendrian isotopy class of Λ.

Theorem 3.1 (Bourgeois, Ekholm, Eliashberg [1]). *Let (X_0, ω_0) be a symplectic manifold with convex boundary (Y_0, ξ_0) such that $C\mathbb{H}(Y_0, \xi_0)$ is defined. Let Λ be a Legendrian sphere in (Y_0, ξ_0) and let (X, ω) be the result of a critical handle attachment on (X_0, ω_0) along Λ. Then we have an exact triangle*

$$\begin{array}{ccc} C\mathbb{H}(X,\omega) & \xrightarrow{\quad [0] \quad} & C\mathbb{H}(X_0,\omega_0) \\ {\scriptstyle [0]} \nwarrow & & \swarrow {\scriptstyle [-1]} \\ & L\mathbb{H}^{cyc}(\Lambda) & \end{array}$$

In the above Theorem, the map $C\mathbb{H}(X, \omega) \to C\mathbb{H}(X_0, \omega_0)$ counts rigid capped holomorphic curves in the symplectic cobordism $X \setminus X_0$. The map $C\mathbb{H}(X_0, \omega_0) \to L\mathbb{H}^{\mathrm{cyc}}(\Lambda)$ counts rigid capped holomorphic curves in the symplectization $\mathbb{R} \times X_0$ with boundary on $\mathbb{R} \times \Lambda$, that are asymptotic to a closed Reeb orbit at the convex end and to Reeb chords of Λ at the concave end.

Noncyclic version. We now define another variant of the complex $(LH(\Lambda), d_{LH})$, which plays an important role to describe the effect of Legendrian surgery on $S\mathbb{H}^+(X, \omega)$. We define a module $LH^{\mathrm{Ho},+}(\Lambda) = \widecheck{LH}^+(\Lambda) \oplus \widehat{LH}^+(\Lambda)$, where $\widecheck{LH}^+(\Lambda) = LH^+(\Lambda)$ and $\widehat{LH}^+(\Lambda) = LH^+(\Lambda)[1]$. If $w = c_1 \dots c_k$ is a word of Reeb chords for Λ in $LH^+(\Lambda)$ we denote by $\widecheck{w} = \widecheck{c}_1 c_2 \dots c_k$ and by $\widehat{w} = \widehat{c}_1 c_2 \dots c_k$ the corresponding generators in $\widecheck{LH}^+(\Lambda)$ and in $\widehat{LH}^+(\Lambda)$ respectively. In what follows, we will sometimes write a word with a decoration (check or hat) which is not in the first position. As a convention, we identify such a word with its cyclic permutation starting with the decorated chord, multiplied with a sign as in the definition of P, where $|c| = |\widecheck{c}| = |\widehat{c}| - 1$.

We denote by $\check{d} : \widecheck{LH}^{+}_{*}(X, \omega) \to \widecheck{LH}^{+}_{*-1}(X, \omega)$ the differential defined by

$$\check{d}\check{w} = (d_{LH} w)\check{}.$$

Let $S : LH(\Lambda) \to \widehat{LH}^{+}(\Lambda)$ be the linear map defined by $S(1) = 0$ and

$$S(c_1 \dots c_k) = \widehat{c_1} c_2 \dots c_k + \dots + (-1)^{|c_1| + |\dots + |c_{k-1}|} c_1 \dots c_{k-1} \widehat{c_k}.$$

We denote by $\widehat{d} : \widehat{LH}^{+}_{*}(X, \omega) \to \widehat{LH}^{+}_{*-1}(X, \omega)$ the differential defined by

$$\widehat{d}(\widehat{c} w') = S(d_{LH} c) w' + (-1)^{|c|+1} \widehat{c} (d_{LH} w').$$

We define a linear map $\alpha : \widehat{LH}^{+}_{*}(\Lambda) \to \widecheck{LH}^{+}_{*-1}(\Lambda)$ by

$$\alpha(\widehat{c_1} c_2 \dots c_k) = \check{c_1} c_2 \dots c_k - c_1 \dots c_{k-1} \check{c_k}.$$

The **differential** $d_{\mathrm{Ho},+} : LH^{\mathrm{Ho},+}_{*}(\Lambda) \to LH^{\mathrm{Ho},+}_{*-1}(\Lambda)$ is defined, with respect to the above decomposition of $LH^{\mathrm{Ho},+}$, by

$$d_{\mathrm{Ho},+} = \begin{pmatrix} \check{d} & \alpha \\ 0 & \widehat{d} \end{pmatrix}.$$

Then $d_{\mathrm{Ho},+} \circ d_{\mathrm{Ho},+} = 0$ and the homology $L\mathbb{H}^{\mathrm{Ho},+}(\Lambda) = H(LH^{\mathrm{Ho},+}(\Lambda), d_{\mathrm{Ho},+})$ of the corresponding complex is again an invariant of the Legendrian isotopy class of Λ.

Note that, with the above algebraic definitions, we have an exact triangle

$$\begin{array}{ccc} L\mathbb{H}^{\mathrm{cyc}}(\Lambda) & \longrightarrow & L\mathbb{H}^{\mathrm{cyc}}(\Lambda) \\ \nwarrow & & \swarrow \\ & L\mathbb{H}^{\mathrm{Ho},+}(\Lambda) & \end{array}$$

which is similar to the Connes exact triangle for cyclic and Hochschild homologies. The analogy between the above exact triangle and the exact triangle in Theorem 2.1 motivates the following result.

Theorem 3.2 (Bourgeois, Ekholm, Eliashberg [1]). *Under the same assumptions as in Theorem 3.1, we have an exact triangle*

$$\begin{array}{ccc} S\mathbb{H}^{+}(X, \omega) & \xrightarrow{[0]} & S\mathbb{H}^{+}(X_0, \omega_0) \\ {\scriptstyle [0]}\nwarrow & & \swarrow{\scriptstyle [-1]} \\ & L\mathbb{H}^{Ho,+}(\Lambda) & \end{array}$$

Expanded version. We finally define another variant of the complex $(LH(\Lambda), d_{LH})$, which plays an important role to describe the effect of Legendrian surgery on $S\mathbb{H}(X, \omega)$. We define an expanded module $LH^{\mathrm{Ho}}(\Lambda) = LH^{\mathrm{Ho},+}(\Lambda) \oplus \langle \tau \rangle$. The unique generator for the second summand formally represents the unique connected component of Λ.

We define a linear map $T : LH^{\mathrm{Ho},+}(\Lambda) \to \langle \tau \rangle$ by

$$\begin{aligned} T(\check{c}) &= \#\mathcal{M}(c)/\mathbb{R}\, \tau, \\ T(\check{w}) &= 0 \text{ if } w \text{ contains more than one letter}, \\ T(\widehat{w}) &= 0. \end{aligned}$$

In this definition, the moduli space $\mathcal{M}(c)$ consists of holomorphic disks in $\mathbb{R} \times Y$ with boundary on $\mathbb{R} \times \Lambda$, having one boundary puncture converging at the convex end of $\mathbb{R} \times Y$ to the Reeb chord c.

The **differential** $d_{\mathrm{Ho}} : LH^{\mathrm{Ho}}_*(\Lambda) \to LH^{\mathrm{Ho}}_{*-1}(\Lambda)$ is defined, with respect to the above decomposition of LH^{Ho}, by

$$d_{\mathrm{Ho}} = \begin{pmatrix} d_{\mathrm{Ho},+} & 0 \\ T & 0 \end{pmatrix}.$$

Then $d_{\mathrm{Ho}} \circ d_{\mathrm{Ho}} = 0$ and the corresponding homology $L\mathbb{H}^{\mathrm{Ho}}(\Lambda) = H(LH^{\mathrm{Ho}}(\Lambda), d_{\mathrm{Ho}})$ is again an invariant of the Legendrian isotopy class of Λ.

Theorem 3.3 (Bourgeois, Ekholm, Eliashberg [1]). *Under the same assumptions as in Theorem 3.1, we have an exact triangle*

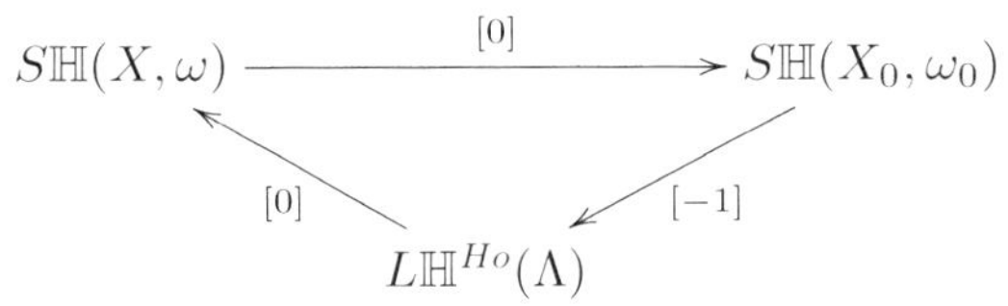

In this last result, note that for a subcritical Stein manifold (X_0, ω_0), since we have $S\mathbb{H}(X_0, \omega_0) = 0$, it follows that

$$S\mathbb{H}(X, \omega) \simeq L\mathbb{H}^{\mathrm{Ho}}(\Lambda).$$

References

[1] F. Bourgeois, T. Ekholm and Y. Eliashberg, Effect of Legendrian surgery, arXiv preprint, arXiv:0911.0026.

[2] F. Bourgeois, Y. Eliashberg, H. Hofer, K. Wysocki and E. Zehnder, Compactness results in Symplectic Field Theory, *Geom. Topol.* **7** (2003), 799–888.

[3] F. Bourgeois and K. Mohnke, Coherent orientations in Symplectic Field Theory, *Math. Z.* **248** (2004) no. 1, 123–146.

[4] F. Bourgeois and A. Oancea, An exact sequence for contact- and symplectic homology, *Invent. Math.* **175** (2009) no. 3, 611–680.

[5] F. Bourgeois and A. Oancea, The Gysin exact sequence for S^1-equivariant symplectic homology, arXiv preprint, arXiv:0909.4526.

[6] F. Bourgeois, A. Oancea, Linearized contact homology and S^1-equivariant symplectic homology, *in preparation.*

[7] K. Cieliebak, Handle attaching in symplectic homology and the chord conjecture, *J. Eur. Math. Soc.* **4** (2002), no. 2, 115–142.

[8] T. Ekholm, J. Etnyre and M. Sullivan, The contact homology of Legendrian submanifolds in $\mathbb{R}^{2n+1}$, *J. Differential Geom.* **71** (2005), no. 2, 177–305.

[9] T. Ekholm, J. Etnyre and M. Sullivan, Legendrian contact homology in $P \times \mathbb{R}$, *Trans. Amer. Math. Soc.* **359** (2007), no. 7, 3301–3335.
[10] Y. Eliashberg, Topological characterization of Stein manifolds of dimension > 2, *Internat. J. Math.* **1** (1990), no. 1, 29–46.
[11] Y. Eliashberg, A. Givental and H. Hofer, Introduction to Symplectic Field Theory, *Geom. Funct. Anal.*, Special Volume, Part II (2000), 560–673.
[12] A. Floer, The unregularized gradient flow of the symplectic action, *Comm. Pure Appl. Math.* **41** (1988), 393–407.
[13] H. Grauert, On Levi's problem and the imbedding of real-analytic manifolds, *Ann. of Math.* (2) **68** (1958), 460–472.
[14] M. Gromov, Pseudo-holomorphic Curves in Symplectic Manifolds, *Invent. Math.* **82** (1985), 307–347.
[15] H. Hofer, K. Wysocki and E. Zehnder, A general Fredholm theory. I. A splicing-based differential geometry, *J. Eur. Math. Soc.* **9** (2007), no. 4, 841–876.
[16] H. Hofer, K. Wysocki and E. Zehnder, A general Fredholm theory. II. Implicit function theorems, *Geom. Funct. Anal.* **19** (2009), no. 1, 206–293.
[17] H. Hofer, K. Wysocki and E. Zehnder, A general Fredholm theory. III. Fredholm functors and polyfolds, *Geom. Topol.* **13** (2009), no. 4, 2279–2387.
[18] A. Oancea, A survey of Floer homology for manifolds with contact type boundary or symplectic homology, *Ensaios Mat.* **7** (2004), 51–91.
[19] J. Robbin and D. Salamon, The Maslov index for paths, *Topology* **32** (1993), No 4, 827–844.
[20] C. Viterbo, Functors and computations in Floer homology with applications. I. *Geom. Funct. Anal.* **9** (1999), no. 5, 985–1033.
[21] M.-L. Yau, Cylindrical contact homology of subcritical Stein-fillable contact manifolds, *Geom. Topol.* **8** (2004), 1243–1280.
[22] A. Weinstein, Contact surgery and symplectic handlebodies, *Hokkaido Math. J.* **20** (1991), no. 2, 241–251.

Département de Mathématique CP218, Université Libre de Bruxelles, Boulevard du Triomphe, 1050 Bruxelles, Belgium

E-mail address: fbourgeo@ulb.ac.be

Proceedings of 17th Gökova
Geometry-Topology Conference
pp. 21 – 63

A solution to the Arf-Kervaire invariant problem

Michael A. Hill, Michael J. Hopkins and Douglas C. Ravenel

Abstract. This paper gives the history and background of one of the oldest problems in algebraic topology, along with an outline of our solution to it. A rigorous account can be found in our preprint [HHR]. The third author has a website with numerous links to related papers and talks we have given on the subject since announcing our result in April, 2009.

Drawing by Carolyn Snaith, 1981

Partially supported by DARPA and NSF.

Contents

Main Theorem. *The Arf-Kervaire elements* $\theta_j \in \pi^S_{2^{j+1}-2}$ *do not exist for* $j \geq 7$.

Here π_k^S denotes the kth stable homotopy group of spheres, which will be defined shortly.

The kth (for a positive integer k) homotopy group of the topological space X, denoted by $\pi_k(X)$, is the set of continuous maps to X from the k-sphere S^k, up to continuous deformation. For technical reasons we require that each map send a specified point in S^k (called a *base point*) to a specified point $x_0 \in X$. When X is path connected the choice of these two points is irrelevant, so it is usually omitted from the notation. When X is not path connected, we get different collections of maps depending on the path connected component of the base point.

This set has a natural group structure, which is abelian for $k > 1$. The word *natural* here means that a continuous base point preserving map $f : X \to Y$ induces a homomorphism $f_* : \pi_k X \to \pi_k Y$, sometimes denoted by $\pi_k f$.

It is known that the group $\pi_{n+k}S^n$ is independent of n for $n > k$. There is a homomorphism $E : \pi_{n+k}S^n \to \pi_{n+k+1}S^{n+1}$ defined as follows. S^{n+1} [respectively S^{n+k+1}] can be obtained from S^n [S^{n+k}] by a double cone construction known as *suspension*. The cone over S^n is an $(n+1)$-dimensional ball, and gluing two such balls together along their common boundary gives an $(n+1)$-dimensional sphere. A map $f : S^{n+k} \to S^n$ can be canonically extended (by suspending both its source and target) to a map $Ef : S^{n+k+1} \to S^{n+1}$, and this leads to the suspension homomorphism E. The Freudenthal Suspension Theorem [Fre38], proved in 1938, says that it is onto for $k = n$ and an isomorphism for $n > k$. For this reason the group $\pi_{n+k}S^n$ is said to be *stable* when $n > k$, and it is denoted by π_k^S and called the *stable k-stem*.

The Main Theorem above concerns the case $k = 2^{j+1} - 2$. The θ_j in the theorem is a hypothetical element related to a geometric invariant of certain manifolds studied originally by Pontryagin starting in the 1930s, [Pon38], [Pon50] and [Pon55]. The problem came into its present form with a theorem of Browder [Bro69] published in 1969. There were several unsuccessful attempts to solve it in the 1970s. They were all aimed at proving the opposite of what we have proved, namely that all of the θ_j exist.

The θ_j in the theorem is the name given to a hypothetical map between spheres for which the Arf-Kervaire invariant is nontrivial. Browder's theorem says that such things can exist only in dimensions that are 2 less than a power of 2.

Some homotopy theorists, most notably Mahowald, speculated about what would happen if θ_j existed for all j. They derived numerous consequences about homotopy groups of spheres. The possible nonexistence of the θ_j for large j was known as the DOOMSDAY HYPOTHESIS.

After 1980, the problem faded into the background because it was thought to be too hard. In 2009, just a few weeks before we announced our theorem, Snaith published a book [Sna09] on the problem "to stem the tide of oblivion." On the difficulty of the problem, he wrote

> In the light of ... the failure over fifty years to construct framed manifolds of Arf-Kervaire invariant one this might turn out to be a book about things which do not exist. This [is] why the quotations which preface each chapter contain a preponderance of utterances from the pen of Lewis Carroll.

Our proof is two giant steps away from anything that was attempted in the 70s. We now know that the world of homotopy theory is very different from what they had envisioned then.

1. Background and history

1.1. Pontryagin's early work on homotopy groups of spheres

The Arf-Kervaire invariant problem has its origins in Pontryagin's early work on a geometric approach to the homotopy groups of spheres, [Pon38], [Pon50] and [Pon55].

Pontryagin's approach to maps $f : S^{n+k} \to S^n$ is to assume that f is smooth and that the base point y_0 of the target is a regular value. (Any continuous f can be continuously deformed to a map with this property.) This means that $f^{-1}(y_0)$ is a closed smooth k-manifold M in S^{n+k}. Let D^n be the closure of an open ball around y_0. If it is sufficiently small, then $V^{n+k} = f^{-1}(D^n)$ in S^{n+k} is an $(n+k)$-manifold homeomorphic to $M \times D^n$ with boundary homeomorphic to $M \times S^{n-1}$. It is also a tubular neighborhood of M^k and comes equipped with a map $p : V^{n+k} \to M^k$ sending each point to the nearest point in M. For each $x \in M$, $p^{-1}(x)$ is homeomorphic to a closed n-ball B^n. The pair $(p, f|V^{n+k})$ defines an explicit homeomorphism

$$V^{n+k} \xrightarrow[\approx]{(p,f|V^{n+k})} M^k \times D^n.$$

This structure on M^k is called a *framing*, and M is said to be *framed in* $\mathbf{R}^{n+k}$. A choice of basis of the tangent space at $y_0 \in S^n$ pulls back to a set of linearly independent normal vector fields on $M \subset \mathbf{R}^{n+k}$.

Conversely, suppose we have a closed sub-k-manifold $M \subset \mathbf{R}^{n+k}$ with a closed tubular neighborhood V and a homeomorphism h to $M \times D^n$ as above. This is called a *framed sub-k-manifold* of $\mathbf{R}^{n+k}$. Some remarks are in order here.

- The existence of a framing puts some restrictions on the topology of M. All of its characteristic classes must vanish. In particular it must be orientable.
- A framing can be twisted by a map $g : M \to SO(n)$, where $SO(n)$ denotes the group of orthogonal $n \times n$ matrices with determinant 1. Such matrices act on D^n in an obvious way. The twisted framing is the composite

$$\begin{array}{ccccc} V & \xrightarrow{h} & M^k \times D^n & \longrightarrow & M^k \times D^n \\ & & (m, x) & \longmapsto & (m, g(m)(x)). \end{array}$$

  We will say more about this later.

- If we drop the assumption that M is framed, then the tubular neighborhood V is a (possibly nontrivial) disk bundle over M. The map $M \to y_0$ needs to be replaced by a map to the classifying space for such bundles, $BO(n)$. This leads to unoriented bordism theory, which was analyzed by Thom in [Tho54]. Two helpful references for this material are the books by Milnor-Stasheff [MS74] and Stong [Sto68].

Pontryagin constructs a map $P(M,h) : S^{n+k} \to S^n$ as follows. We regard S^{n+k} as the one point compactification of $\mathbf{R}^{n+k}$ and S^n as the quotient $D^n/\partial D^n$. This leads to a diagram

$$\begin{array}{ccccc}
(V, \partial V) & \xrightarrow{\quad h \quad} & M \times (D^n, \partial D^n) & \xrightarrow{\quad p_2 \quad} & (D^n, \partial D^n) \\
\big\downarrow & & & & \big\downarrow \\
(\mathbf{R}^{n+k}, \mathbf{R}^{n+k} - \mathrm{int} V) & \longrightarrow & (S^{n+k}, S^{n+k} - \mathrm{int} V) & \xrightarrow{P(M,h)} & (S^n, \{\infty\}) \\
 & & S^{n+k} - \mathrm{int} V & \xmapsto{\quad P(M,h) \quad} & \{\infty\}
\end{array}$$

The map $P(M,h)$ is the extension of $p_2 h$ obtained by sending the complement of V in S^{n+k} to the point at infinity in S^n. For $n > k$, the choice of the embedding (but not the choice of framing) of M into the Euclidean space is irrelevant. Any two embeddings (with suitably chosen framings) lead to the same map $P(M,h)$ up to continuous deformation.

To proceed further, we need to be more precise about what we mean by continuous deformation. Two maps $f_1, f_2 : X \to Y$ are *homotopic* if there is a continuous map $h : X \times [0,1] \to Y$ (called a *homotopy between* f_1 *and* f_2) such that

$$h(x,0) = f_1(x) \qquad \text{and} \qquad h(x,1) = f_2(x).$$

Now suppose $X = S^{n+k}$, $Y = S^n$, and the map h (and hence f_1 and f_2) is smooth with y_0 as a regular value. Then $h^{-1}(y_0)$ is a framed $(k+1)$-manifold N whose boundary is the disjoint union of $M_1 = f_1^{-1}(y_0)$ and $M_2 = f_2^{-1}(y_0)$. This N is called a *framed cobordism* between M_1 and M_2, and when it exists the two closed manifolds are said to be *framed cobordant.*

Let $\Omega^{\mathrm{fr}}_{k,n}$ denote the cobordism group of framed k-manifolds in $\mathbf{R}^{n+k}$. The above construction leads to Pontryagin's isomorphism

$$\Omega^{\mathrm{fr}}_{k,n} \xrightarrow{\approx} \pi_{n+k} S^n.$$

First consider the case $k = 0$. Here the 0-dimensional manifold M is a finite set of points in $\mathbf{R}^n$. Each comes with a framing which can be obtained from a standard one by an element in the orthogonal group $O(n)$. We attach a sign to each point corresponding to the sign of the associated determinant. With these signs we can count the points algebraically and get an integer called the *degree of* f. Two framed 0-manifolds are cobordant iff they have the same degree.

Sidebar 1 The Hopf-Whitehead J-homomorphism

Suppose our framed manifold is S^k with a framing that extends to a D^{k+1}. This will lead to the trivial element in $\pi_{n+k}(S^n)$, but twisting the framing can lead to nontrivial elements. The twist is determined up to homotopy by an element in $\pi_k(SO(n))$. Pontryagin's construction thus leads to the homomorphism

$$\pi_k SO(n) \xrightarrow{J} \pi_{n+k} S^n$$

introduced by Hopf [Hop35] and Whitehead [Whi42]. Both source and target are known to be independent of n for $n > k+1$. In this case the source group for each k (denoted simply by $\pi_k SO$ since n is irrelevant) was determined by Bott [Bot59] in his remarkable periodicity theorem. He showed

$$\pi_k SO = \begin{cases} \mathbf{Z} & \text{for } k \equiv 3 \text{ or } 7 \bmod 8 \\ \mathbf{Z}/2 & \text{for } k \equiv 0 \text{ or } 1 \bmod 8 \\ 0 & \text{otherwise.} \end{cases}$$

Here is a table showing these groups for $k \leq 10$.

k	1	2	3	4	5	6	7	8	9	10
$\pi_k(SO)$	$\mathbf{Z}/2$	0	$\mathbf{Z}$	0	0	0	$\mathbf{Z}$	$\mathbf{Z}/2$	$\mathbf{Z}/2$	0

In each case where the group is nontrivial, the image under J of its generator is known to generate a direct summand. In the jth case we denote this image by β_j and its dimension by $\phi(j)$, which is roughly $2j$. The first three of these are the Hopf maps $\eta \in \pi_1^S$, $\nu \in \pi_3^S$ and $\sigma \in \pi_7^S$. After that we have $\beta_4 \in \pi_8^S$, $\beta_5 \in \pi_9^S$, $\beta_6 \in \pi_{11}^S$ and so on. For the case $\pi_{4m-1} SO = \mathbf{Z}$, the image under J is known to be a cyclic group whose order a_m is the denominator of $B_m/4m$, where B_m is the mth Bernoulli number. Details can be found in [Ada66] and [MS74]. Here is a table showing these values for $m \leq 10$.

m	1	2	3	4	5	6	7	8	9	10
a_m	24	240	504	480	264	65,520	24	16,320	28,728	13,200

Now consider the case $k = 1$. M is a closed 1-manifold, i.e., a disjoint union of circles. Two framings on a single circle differ by a map from S^1 to the group $SO(n)$, and it is known that

$$\pi_1 SO(n) = \begin{cases} 0 & \text{for } n = 1 \\ \mathbf{Z} & \text{for } n = 2 \\ \mathbf{Z}/2 & \text{for } n > 2. \end{cases}$$

It turns out that any disjoint union of framed circles is cobordant to a single framed circle. This can be used to show that

$$\pi_{n+1}S^n = \begin{cases} 0 & \text{for } n = 1 \\ \mathbf{Z} & \text{for } n = 2 \\ \mathbf{Z}/2 & \text{for } n > 2. \end{cases}$$

The case $k = 2$ is more subtle. As in the 1-dimensional case we have a complete classification of closed 2-manifolds, and it is only necessary to consider ones which are path connected and oriented. The existence of a framing implies that the surface is orientable, so it is characterized by its genus.

If the genus is zero, namely if $M = S^2$, then there is a framing which extends to a 3-dimensional ball. This makes M cobordant to the empty set, which means that the map is *null homotopic* (or, more briefly, *null*), meaning that it is homotopic to a constant map. Any two framings on S^2 differ by an element in $\pi_2 SO(n)$. This group is known to vanish (as is π_2 of any Lie group), so any two framings on S^2 are equivalent, and the map $f : S^{n+2} \to S^n$ is null.

Now suppose the genus is one and suppose we can find an embedded arc on which the framing extends to a disk in S^{n+2}. Then there is a cobordism which effectively cuts along the arc, leaving a surface bounded by two circles, and attaches two framed disks to them. This process is called *framed surgery.* If we can do this, then we have converted the torus to a 2-sphere and shown that the map $f : S^{n+2} \to S^n$ is null.

When can we find such a closed curve in M? It must represent a generator of $H_1(M)$ and carry a trivial framing. This leads to a map

$$\varphi : H_1(M; \mathbf{Z}/2) \to \mathbf{Z}/2 \tag{1}$$

defined as follows. Each class in H_1 can be represented by a closed curve which is framed either trivially or nontrivially. It can be shown that homologous curves have the same framing invariant, so φ is well defined. At this point Pontryagin made a famous mistake which went undetected for over a decade: he assumed that φ was a homomorphism. We now know this is not the case, and we will say more about it below in §1.3.

On that basis he argued that φ must have a nontrivial kernel, since the source group is $(\mathbf{Z}/2)^2$. Therefore there is a closed curve along which we can do the surgery. It follows that M can be surgered into a 2-sphere, leading to the erroneous conclusion that $\pi_{n+2}(S^n) = 0$ for all n. Freudenthal [Fre38] and later George Whitehead [Whi50] both proved that it is $\mathbf{Z}/2$ for $n \geq 2$. Pontryagin corrected his mistake in [Pon50], and in [Pon55] he gave a complete account of the relation between framed cobordism and homotopy groups of spheres.

1.2. Our main result

Our main theorem can be stated in three different but equivalent ways:

- *Manifold formulation:* It says that a certain geometrically defined invariant $\Phi(M)$ (the Arf-Kervaire invariant, to be defined later) on certain manifolds M is always zero.
- *Stable homotopy theoretic formulation:* It says that certain long sought hypothetical maps between high dimensional spheres do not exist.
- *Unstable homotopy theoretic formulation:* It says something about the EHP sequence (to be defined below), which has to do with unstable homotopy groups of spheres.

Here again is the stable homotopy theoretic formulation.

Main Theorem. *The Arf-Kervaire elements $\theta_j \in \pi^S_{2^{j+1}-2}$ do not exist for $j \geq 7$.*

1.3. The manifold formulation

Let λ be a nonsingular anti-symmetric bilinear form on a free abelian group H of rank $2n$ with mod 2 reduction $\overline{H}$. It is known that $\overline{H}$ has a basis of the form $\{a_i, b_i \colon 1 \leq i \leq n\}$ with

$$\lambda(a_i, a_{i'}) = 0 \qquad \lambda(b_j, b_{j'}) = 0 \qquad \text{and} \qquad \lambda(a_i, b_j) = \delta_{i,j}.$$

In other words, $\overline{H}$ has a basis for which the bilinear form's matrix has the symplectic form

$$\begin{bmatrix} 0 & 1 & & & & & \\ 1 & 0 & & & & & \\ & & 0 & 1 & & & \\ & & 1 & 0 & & & \\ & & & & \ddots & & \\ & & & & & 0 & 1 \\ & & & & & 1 & 0 \end{bmatrix}.$$

A *quadratic refinement of* λ is a map $q : \overline{H} \to \mathbf{Z}/2$ satisfying

$$q(x+y) = q(x) + q(y) + \lambda(x,y).$$

Its *Arf invariant* is

$$\operatorname{Arf}(q) = \sum_{i=1}^{n} q(a_i)q(b_i) \in \mathbf{Z}/2.$$

In 1941 Arf [Arf41] proved that this invariant (along with the number n) determines the isomorphism type of q.

An equivalent definition is the "democratic invariant" of Browder [Bro69]. The elements of $\overline{H}$ "vote" for either 0 or 1 by the function q. The winner of the election (which is never a tie) is $\operatorname{Arf}(q)$. Here is a table illustrating this for three possible refinements q, q' and q'' when $\overline{H}$ has rank 2.

x	0	a	b	$a+b$	Arf invariant
$q(x)$	0	0	0	1	0
$q'(x)$	0	1	1	1	1
$q''(x)$	0	1	0	0	0

The value each refinement on $a+b$ is determined by those on a and b, and q'' is isomorphic to q. Thus the vote is three to one in each case. When $\overline{H}$ has rank 4, it is 10 to 6.

Let M be a $2m$-connected smooth closed manifold of dimension $4m+2$ with a framed embedding in $\mathbf{R}^{4m+2+n}$. We saw above that this leads to a map $f: S^{n+4m+2} \to S^n$ and hence an element in $\pi_{n+4m+2}S^n$.

Let $H = H_{2m+1}(M;\mathbf{Z})$, the homology group in the middle dimension. Each $x \in H$ is represented by an immersion $i_x : S^{2m+1} \looparrowright M$ with a stably trivialized normal bundle. H has an antisymmetric bilinear form λ defined in terms of intersection numbers.

In 1960 Kervaire [Ker60] defined a quadratic refinement q on its mod 2 reduction in terms of the trivialization of each sphere's normal bundle. The *Kervaire invariant* $\Phi(M)$ is defined to be the Arf invariant of q. In the case $m = 0$, when the dimension of the manifold is 2, Kervaire's q is Pontryagin's map φ of (1).

What can we say about $\Phi(M)$?

- Kervaire [Ker60] showed it must vanish when $k = 2$. This enabled him to construct the first example of a topological manifold (of dimension 10) without a smooth structure. Let N be a smooth 10-manifold with boundary given as the union of two copies of the tangent disk bundle of S^5, glued together along a common copy of $D^5 \times D^5$ where the fibers in one copy are parallel to the base in the other. The boundary is homeomorphic to S^9. Thus we can get a closed topological manifold X by gluing on a 10-ball along its common boundary with n, or equivalently collapsing ∂N to a point. X then has nontrivial Kervaire invariant. On the other hand, Kervaire proved that any smooth framed manifold must have trivial Kervaire invariant. Therefore the topological framed manifold X cannot have a smooth structure. Equivalently, the boundary ∂N cannot be diffeomorphic to S^9. It must be an exotic 9-sphere.
- For $k = 0$ there is a framing on the torus $S^1 \times S^1 \subset \mathbf{R}^4$ with nontrivial Kervaire invariant. Pontryagin used it in [Pon50] (after some false starts in the 30s) to show $\pi_{n+2}S^n = \mathbf{Z}/2$ for all $n \geq 2$.
- There are similar constructions for $k = 1$ and $k = 3$, where the framed manifolds are $S^3 \times S^3$ and $S^7 \times S^7$ respectively. Like S^1, S^3 and S^7 are both parallelizable, meaning that their tangent bundles are trivial. The framings can be twisted in such a way as to yield a nontrivial Kervaire invariant.
- Brown-Peterson [BP66b] showed that it vanishes for all positive even k. This means that apart from the 2-dimensional case, any smooth framed manifold with nontrivial Kervaire invariant must have a dimension congruent to 6 modulo 8.
- Browder [Bro69] showed that it can be nontrivial only if $k = 2^{j-1} - 1$ for some positive integer j. This happens iff the element h_j^2 is a permanent cycle in the

Adams spectral sequence, which was originally introduced in [Ada58]. (More information about it can be found in [Rav86] and [Rav04].) The corresponding element in $\pi^S_{n+2^{j+1}-2}$ is θ_j, the subject of our theorem. *This is the stable homotopy theoretic formulation of the problem.*

- θ_j is known to exist for $1 \leq j \leq 3$, i.e., in dimensions 2, 6, and 14. In these cases the relevant framed manifold is $S^{2^j-1} \times S^{2^j-1}$ with a twisted framing as discussed above. The framings on S^{2^j-1} represent the elements h_j in the Adams spectral sequence. The Hopf invariant one theorem of Adams [Ada60] says that for $j > 3$, h_j is not a permanent cycle in the Adams spectral sequence because it supports a nontrivial differential. (His original proof was not written in this language, but had to do with secondary cohomology operations.) This means that for $j > 3$, a smooth framed manifold representing θ_j (i.e., having a nontrivial Kervaire invariant) cannot have the form $S^{2^j-1} \times S^{2^j-1}$.
- θ_j is also known to exist for $j = 4$ and $j = 5$, i.e., in dimensions 30 and 62. In both cases the existence was first established by purely homotopy theoretic means, without constructing a suitable framed manifold. For $j = 4$ this was done by Barratt, Mahowald and Tangora in [MT67] and [BMT70]. A framed 30-manifold with nontrivial Kervaire invariant was later constructed by Jones [Jon78]. For $j = 5$ the homotopy theory was done in 1985 by Barratt-Jones-Mahowald in [BJM84].
- Our theorem says θ_j does *not* exist for $j \geq 7$. The case $j = 6$ is still open.

Kervaire's 10-manifold with boundary that is described above can be generalized to a $(4k+2)$-manifold with boundary constructed in a similar way. In all cases except $k = 0$, 1 or 3, any framing of this manifold will do because the tangent bundle of S^{2k+1} is nontrivial and leads to a nontrivial invariant. The boundary is homeomorphic to S^{4k+1}, but may or may not be diffeomorphic to the standard sphere. If it is, then attaching a $(4k+2)$-disk to it will produce a smooth framed manifold with nontrivial Kervaire invariant. If it is not, then we have an exotic $(4k+1)$-sphere bounding a framed manifold and hence not detected by framed cobordism.

1.4. The unstable formulation

Assume all spaces in sight are localized at the prime 2. For each $n > 0$ there is a fiber sequence due to James, [Jam55], [Jam56a], [Jam56b] and [Jam57]

$$S^n \xrightarrow{E} \Omega S^{n+1} \xrightarrow{H} \Omega S^{2n+1}.$$

Here $\Omega X = \Omega^1 X$ where $\Omega^k X$ denotes the space of continuous base point preserving maps to X from the k-sphere S^k, known as the kth loop space of X. This leads to a long exact sequence of homotopy groups

$$\cdots \longrightarrow \pi_{m+n} S^n \xrightarrow{E} \pi_{m+n+1} S^{n+1} \xrightarrow{H} \pi_{m+n+1} S^{2n+1} \xrightarrow{P} \pi_{m+n-1} S^n \longrightarrow \cdots$$

Here

- E stands for *E*inhängung, the German word for suspension.
- H stands for *H*opf invariant.
- P stands for Whitehead *p*roduct.

Assembling these for fixed m and various n leads to a diagram

$$
\begin{array}{ccccccccc}
 & & \pi_{m+n+1}S^{2n-1} & & \pi_{m+n+2}S^{2n+1} & & \pi_{m+n+3}S^{2n+3} & & \\
 & & \downarrow{\scriptstyle P} & & \downarrow{\scriptstyle P} & & \downarrow{\scriptstyle P} & & \\
\cdots & \xrightarrow{E} & \pi_{m+n-1}S^{n-1} & \xrightarrow{E} & \pi_{m+n}S^{n} & \xrightarrow{E} & \pi_{m+n+1}S^{n+1} & \xrightarrow{E} & \cdots \\
 & & \downarrow{\scriptstyle H} & & \downarrow{\scriptstyle H} & & \downarrow{\scriptstyle H} & & \\
 & & \pi_{m+n-1}(S^{2n-3}) & & \pi_{m+n}S^{2n-1} & & \pi_{m+n+1}S^{2n+1} & &
\end{array}
$$

where

- Sequences of arrows labeled H, P, E, H, P (or any subset thereof) in that order are exact.
- The groups in the top and bottom rows are inductively known, and we can compute those in the middle row by induction on n.
- The groups in the top and bottom rows vanish for large n, making E an isomorphism.
- An element in the middle row has trivial suspension (is killed by E) iff it is in the image of P.
- It desuspends (is in the image of E) iff its Hopf invariant (image under H) is trivial.

When $m = n - 1$ this diagram is

$$
\begin{array}{ccccccccc}
 & & & & \pi_{2n+1}S^{n+1} & & & & \\
 & & & & \downarrow{\scriptstyle H} & & & & \\
 & & \pi_{2n}S^{2n-1} & & \pi_{2n+1}S^{2n+1} = \mathbf{Z} & & \pi_{2n+2}S^{2n+3} = 0 & & \\
 & & \downarrow{\scriptstyle P} & & \downarrow{\scriptstyle P} & & \downarrow{\scriptstyle P} & & \\
\cdots & \xrightarrow{E} & \pi_{2n-2}S^{n-1} & \xrightarrow{E} & \pi_{2n-1}S^{n} & \xrightarrow{E} & \pi_{2n}S^{n+1} & \xrightarrow{E} & \cdots \\
 & & \downarrow{\scriptstyle H} & & \downarrow{\scriptstyle H} & & \downarrow{\scriptstyle H} & & \\
 & & \pi_{2n-2}S^{2n-3} & & \pi_{2n-1}S^{2n-1} = \mathbf{Z} & & \pi_{2n}S^{2n+1} = 0 & &
\end{array}
$$

The image under P of the generator of $\pi_{2n+1}S^{2n+1}$ is denoted by $w_n \in \pi_{2n-1}S^n$ and is called the *Whitehead square*.

- When n is even, $H(w_n) = 2$ and w_n has infinite order.
- w_n is trivial for $n = 1$, 3 and 7. In those cases the generator of $\pi_{2n+1}S^{2n+1}$ is the Hopf invariant (image under H) of one of the three Hopf maps in $\pi_{2n+1}S^{n+1}$,

$$S^3 \xrightarrow{\eta} S^2, \qquad S^7 \xrightarrow{\nu} S^4 \quad \text{and} \qquad S^{15} \xrightarrow{\sigma} S^8.$$

- For other odd values of n, twice the generator of the upper $\mathbf{Z}$ is $H(w_{n+1})$, so w_n has order 2.
- It turns out that w_n is divisible by 2 iff $n = 2^{j+1} - 1$ and θ_j exists, in which case $w_n = 2\theta_j$.
- Each Whitehead square $w_{2n+1} \in \pi_{4n+1}S^{2n+1}$ (except the cases $n = 0$, 1 and 3) desuspends to a lower sphere until we get an element with a nontrivial Hopf invariant, which is always some β_j (see Sidebar 1, page 26). More precisely we have
$$H(w_{(2s+1)2^j-1}) = \beta_j$$
for each $j > 0$ and $s \geq 0$. This result is essentially Adams' 1962 solution to the vector field problem [Ada62].

Recall the EHP sequence

$$\cdots \longrightarrow \pi_{m+n}S^n \xrightarrow{E} \pi_{m+n+1}S^{n+1} \xrightarrow{H} \pi_{m+n+1}S^{2n+1} \xrightarrow{P} \pi_{m+n-1}S^n \longrightarrow \cdots$$

Given some $\beta_j \in \pi_{\phi(j)+2n+1}S^{2n+1}$ for $\phi(j) < 2n$, one can ask about the Hopf invariant of its image under P, which vanishes when β_j is in the image of H. In most cases the answer is known and is due to Mahowald, [Mah67] and [Mah82]. The remaining cases have to do with θ_j. The answer that he had hoped for is the following, which can be found in [Mah67]. (To our knowledge, Mahowald never referred to this as the World Without End Hypothesis. We chose that term to emphasize its contrast with the Doomsday Hypothesis.)

World Without End Hypothesis (Mahowald 1967).

- *The Arf-Kervaire element* $\theta_j \in \pi^S_{2^{j+1}-2}$ *exists for all* $j > 0$.
- *It desuspends to* $S^{2^{j+1}-1-\phi(j)}$ *and its Hopf invariant is* β_j.
- *Let* $j, s > 0$ *and suppose that* $m = 2^{j+2}(s+1) - 4 - \phi(j)$ *and* $n = 2^{j+1}(s+1) - 2 - \phi(j)$. *Then* $P(\beta_j)$ *has Hopf invariant* θ_j.

This describes the systematic behavior in the EHP sequence of elements related to the image of J, and the θ_j are an essential part of the picture. Because of our theorem, *we now know that this hypothesis is incorrect.*

1.5. Questions raised by our theorem

EHP sequence formulation. The World Without End Hypothesis was the nicest possible statement of its kind given all that was known prior to our theorem. Now we know it cannot be true since θ_j does not exist for $j \geq 7$. *This means the behavior of the indicated elements* $P(\beta_j)$ *for* $j \geq 7$ *is a mystery.*

Adams spectral sequence formulation. We now know that the h_j^2 for $j \geq 7$ are not permanent cycles, so they have to support nontrivial differentials. *We have no idea what their targets are.*

Manifold formulation. Here our result does *not* lead to any obvious new questions. It appears rather to be the final page in the story.

Our method of proof offers a new tool for studying the stable homotopy groups of spheres. We look forward to learning more with it in the future.

2. Our strategy

2.1. Ingredients of the proof

Our proof has several ingredients.

- It uses methods of stable homotopy theory, which means it uses spectra instead of topological spaces. Recall that a space X has a homotopy group $\pi_k X$ for each positive integer k. A spectrum X has an abelian homotopy group $\pi_k X$ *defined for every integer* k. For the sphere spectrum S^0, $\pi_k S^0$ is the stable k-stem homotopy group π_k^S. The hypothetical θ_j is an element of this group for $k = 2^{j+1} - 2$.
- It uses complex cobordism theory, the subject of Sidebar 2. This is a branch of algebraic topology having deep connections with algebraic geometry and number theory. It includes some highly developed computational techniques that began with work by Milnor [Mil60], Novikov ([Nov60], [Nov62] and [Nov67]) and Quillen [Qui69] in the 60s. A pivotal tool in the subject is the theory of formal group laws, the subject of Sidebar 4.
- It also makes use of newer less familiar methods from equivariant stable homotopy theory. A helpful introduction to this subject is the paper of Greenlees-May [GM95]. This means there is a finite group G (a cyclic 2-group) acting on all spaces in sight, and all maps are required to commute with these actions. When we pass to spectra, we get homotopy groups indexed not just by the integers $\mathbf{Z}$, but by $RO(G)$, the orthogonal representation ring of G. Our calculations make use of this richer structure.

2.2. The spectrum Ω

We will produce a map $S^0 \to \Omega$, where Ω is a nonconnective spectrum (meaning that it has nontrivial homotopy groups in arbitrarily large negative dimensions) with the following properties:

(i) *Detection Theorem.* It has an Adams-Novikov spectral sequence (which is a device for calculating homotopy groups) in which the image of each θ_j is nontrivial. *This means that if θ_j exists, we will see its image in $\pi_*\Omega$.*

(ii) *Periodicity Theorem.* It is 256-periodic, meaning that $\pi_k\Omega$ depends only on the reduction of k modulo 256.

(iii) *Gap Theorem.* $\pi_k\Omega = 0$ for $-4 < k < 0$. This property is our zinger. Its proof involves a new tool that we call the slice spectral sequence.

If $\theta_7 \in \pi_{254}S^0$ exists, (i) implies it has a nontrivial image in $\pi_{254}\Omega$. On the other hand, (ii) and (iii) imply that $\pi_{254}\Omega = 0$, so θ_7 cannot exist. The argument for θ_j for larger j is similar, since $|\theta_j| = 2^{j+1} - 2 \equiv -2 \bmod 256$ for $j \geq 7$.

2.3. How we construct Ω

The construction of Ω requires the use of *equivariant stable homotopy theory.* It is discussed extensively in [HHR, §2 and Appendices A and B]. Our spectrum Ω will be the fixed point spectrum for the action of C_8 (the cyclic group of order 8) on an equivariant spectrum $\Omega_{\mathbb{O}}$.

To construct it we start with the complex cobordism spectrum MU. It can be thought of as the set of complex points of an algebraic variety defined over the real numbers. This means that it has an action of C_2 defined by complex conjugation. The resulting C_2-equivariant spectrum is denoted by $MU_{\mathbf{R}}$ and is called *real cobordism theory.* This terminology follows Atiyah's definition of real K-theory [Ati66], by which he meant complex K-theory equipped with complex conjugation.

2.4. The norm

Next we use a formal tool we call the *norm* N_H^G for inducing up from an H-spectrum to a G-spectrum when H is a subgroup of G. It is treated in [HHR, §2.3]. Since the only groups we consider here are cyclic, we will denote N_H^G by N_h^g where $h = |H|$ and $g = |G|$.

The analogous construction on the space level is easy to desribe. Let X be a space that is acted on by a subgroup H of G. Let

$$Y = \mathrm{Map}_H(G, X),$$

the space of H-equivariant maps from G to X. Here the action of H on G is by left multiplication, and the resulting object has an action of G by right multiplication. As a set, $Y = X^{|G/H|}$, the $|G/H|$-fold Cartesian power of X. A general element of G permutes these factors, each of which is left invariant by the subgroup H. For an H-spectrum X, the analogous G-spectrum is denoted by $N_H^G X$.

In particular for a finite cyclic 2-group $G = C_{2^{n+1}}$ we get a G-spectrum

$$MU^{(G)} = N_2^g MU_{\mathbf{R}}$$

underlain by $MU^{(2^n)}$. We are most interested in the case $n = 2$. This spectrum is not periodic, but it has a close relative $\Omega_{\mathbb{O}}$ (to defined below in §5) which is periodic.

Sidebar 2 Complex cobordism

The complex cobordism spectrum MU is a critical tool in modern stable homotopy theory. Its use in computing the stable homotopy groups of spheres is the subject of [Rav86]. Much of the needed background material on vector bundles and Thom spaces can be found in [MS74].

Let $U(n)$ denote the n-dimensional unitary group and $BU(n)$ its classifying space. The latter is the space of complex n-planes in an infinite dimensional complex vector space. It is the base space of an n-dimensional complex vector bundle where the fiber at each point is the set of vectors in the given n-plane. We denote the one point compactification of its total space (also known as the Thom space of the vector bundle) by $MU(n)$. The inclusion map $U(n) \to U(n+1)$ leads to a map from $\Sigma^2 MU(n)$ to $MU(n+1)$, which is adjoint to the map $MU(n) \to \Omega^2 MU(n+1)$.

Such maps can be looped and iterated, allowing us to define

$$MU_k = \lim_{\to} \Omega^{2n-k} MU(n)$$

with MU_k homeomorphic to ΩMU_{k+1}. The collection $\{MU_k\}$ is the spectrum MU. Its homotopy groups are defined by

$$\pi_i MU = \lim_{\to} \pi_{2n+i} MU(n) = \pi_{i+k} MU_k.$$

Its homology and cohomology can be similarly defined. The inclusions of $U(m) \times U(n)$ into $U(m+n)$ lead to maps $MU(m) \wedge MU(n) \to MU(m+n)$ and $MU \wedge MU \to MU$. This makes MU into an E_∞-ring spectrum. There are generalized homology and cohomology theories MU_* and MU^* defined by

$$MU_i(X) = \pi_i X \wedge MU \quad \text{and} \quad MU^i(X) = [X, MU_i],$$

with the latter being endowed with cup products.

We know that

$$\pi_* MU = \mathbf{Z}[x_i \colon i > 0] \quad \text{and} \quad H_* MU = \mathbf{Z}[m_i \colon i > 0]$$

where $|x_i| = |m_i| = 2i$. After localization at a prime p, MU splits into a wedge of copies of a spectrum BP (for Brown-Peterson [BP66a]) with

$$\pi_* BP = \mathbf{Z}_{(p)}[v_i \colon i > 0] \quad \text{and} \quad H_* BP = \mathbf{Z}_{(p)}[\ell_i \colon i > 0]$$

where $|v_i| = |\ell_i| = 2(p^i - 1)$. This is related to a corresponding fact about formal groups laws described in Sidebar 4 (page 51). *In this paper we will assume that MU has been localized at the prime 2.*

2.5. Fixed points and homotopy fixed points

For a G-space X, X^G is the subspace fixed by all of G, which is the same as the space of equivariant maps from a point to X, $\mathrm{Map}_G(*, X)$. To get X^{hG}, we replace the point here by a free contractible G-space EG. The homotopy type of $X^{hG} = \mathrm{Map}_G(EG, X)$ is known to be independent of the choice of EG. The unique map $EG \to *$ leads to a map $\varphi : X^G \to X^{hG}$.

We construct a C_8-spectrum $\Omega_{\mathbb{O}}$ and show that

- $\Omega_{\mathbb{O}}^{hC_8}$ satisfies the detection and Periodicity Theorems.
- $\Omega_{\mathbb{O}}^{C_8}$ satisfies the Gap Theorem.

Hence our proof depends on a fourth property:

(iv) *Fixed Point Theorem.* The map $\Omega_{\mathbb{O}}^{C_8} \to \Omega_{\mathbb{O}}^{hC_8}$ is an equivalence. It proof is given in [HHR, §10]. This fixed point set is our spectrum Ω.

We will come back to the definition of $\Omega_{\mathbb{O}}$ below.

2.6. $RO(G)$-graded homotopy

Let $RO(G)$ denote the orthogonal representation ring of G and let S^V denote the one point compactification of the orthogonal representation V. For a G-space or spectrum X define

$$\pi_V^G X = [S^V, X]_G.$$

Note that when the action of G on V is trivial and V has dimension $|V|$, we denote $\pi_V^G X$ by $\pi_{|V|}^G X$, and an equivariant map $S^V = S^{\dim V} \to X$ must land in X^G, so

$$\pi_n^G X = \pi_n X^G.$$

In the stable category we can make sense of this for virtual as well as actual representations, so we get homotopy groups indexed by $RO(G)$, which we denote collectively by $\pi_\star^G X$. We denote the ordinary homotopy of the underlying spectrum of X by $\pi_*^u X$.

We use the norm in two ways:

- Externally as a functor from H-spectra to G-spectra. In particular, for a representation V of H, $N_h^g S^V = S^{\mathrm{ind} V}$, the one point compactification of the induced representation of V.
- Internally as a homomorphism

  $$N_h^g : \pi_\star^H X \to \pi_\star^G X$$

  for a G-ring spectrum X. Here we are using the forgetful functor to regard X as an H-spectrum with $RO(H)$-graded homotopy. The external norm takes us to $\pi_\star^G N_h^g X$. Then $N_h^g X$ is $X^{(g/h)}$ as a G-spectrum, and we use the multiplication on X to get a map $N_h^g X \to X$.

We will use the same notation for both.

Recall that

$$\pi_* MU = \mathbf{Z}_{(2)}[x_1, x_2, \ldots] \qquad \text{with } |x_i| = 2i.$$

(We are assuming that MU has been localized at 2.) It turns out that any choice of generator $x_i : S^{2i} \to MU$ is the image of the forgetful functor of a map

$$S^{i\rho} \xrightarrow{\overline{x}_i} MU_{\mathbf{R}}.$$

Here ρ denotes the regular real representation of C_2, which is the same thing as the complex numbers $\mathbf{C}$ acted on by conjugation.

For $G = C_{2^{n+1}}$, $\pi_*^u MU^{(G)}$ is a graded polynomial algebra over $\mathbf{Z}_{(2)}$ where

- there are 2^n generators in each positive even dimension $2i$.
- they are acted on transitively by G.

For a group generator $\gamma \in G$ and polynomial generator $r_i \in \pi_{2i}$, the set

$$\left\{\gamma^j r_i : 0 \leq j < 2^n\right\}$$

is algebraically independent, and $\gamma^{2^n} r_i = (-1)^i r_i$.

3. The slice filtration

Now we introduce our main technical tool, which is treated in [HHR, §6]. It is an equivariant analog of the Postnikov tower.

3.1. Postnikov towers

First we need to recall some things about the classical Postnikov towers. The mth Postnikov section $P^m X$ of a space or spectrum X is obtained by killing all homotopy groups of X above dimension m by attaching cells. The fiber of the map $X \to P^m X$ is $P_{m+1}X$, the m-connected cover of X.

These two functors have some universal properties. Let $\mathcal{S}$ and $\mathcal{S}_{>m}$ denote the categories of spectra and m-connected spectra.

The functor P^m is *Farjoun nullification* [Far96] with respect to the subcategory $\mathcal{S}_{>m}$. This means the map $X \to P^m X$ is universal among maps from X to spectra which are $\mathcal{S}_{>m}$*-null* in the sense that all maps to them from m-connected spectra are null. In other words,

- The spectrum $P^m X$ is $\mathcal{S}_{>m}$-null.
- For any $\mathcal{S}_{>m}$-null spectrum Z, the map $\mathcal{S}(P^m X, Z) \to \mathcal{S}(X, Z)$ is an equivalence.

Since $\mathcal{S}_{>m} \subset \mathcal{S}_{>m-1}$, there is a natural transformation $P^m \to P^{m-1}$, whose fiber is denoted by $P^m_m X$. It is an Eilenberg-Mac Lane spectrum with homotopy concentrated in dimension m.

In what follows G will be a finite cyclic 2-group, and $g = |G|$. Let $\mathcal{S}^G$ denote the category of G-equivariant spectra. We need an equivariant analog of $\mathcal{S}_{>m}$. Our choice for this is somewhat novel.

3.2. Slice cells

Recall that $\mathcal{S}_{>m}$, the category of m-connected spectra, is the category of spectra built up out of spheres of dimension $> m$ using arbitrary wedges and mapping cones.

For a subgroup H of G with $|H| = h$ and an integer k, let

$$\widehat{S}(k\rho_H) = G_+ \wedge_H S^{k\rho_H}$$

where ρ_H denotes the regular real representation of H. Its underlying spectrum is a wedge of g/h spheres of dimension kh which are permuted by elements of G and are invariant under H.

We will replace the set of sphere spectra by the collection

$$\mathcal{A} = \left\{\widehat{S}(k\rho_H),\, \Sigma^{-1}\widehat{S}(k\rho_H) \colon H \subset G,\, k \in \mathbf{Z}\right\}.$$

Definition 3.1. *We will refer to the elements in the set $\mathcal{A}$ above as* slice cells *or simply as* cells. *(Note that $\Sigma^{-2}\widehat{S}(k\rho_H)$ and larger desuspensions are not cells.) A* free cell *is one of the form $\widehat{S}(k\rho_{\{e\}})$, a wedge of g k-spheres permuted by G. Note that*

$$\Sigma^{-1}\widehat{S}(k\rho_{\{e\}}) = \widehat{S}((k-1)\rho_{\{e\}}).$$

Nonfree cells are said to be isotropic. *Cells of the form $\widehat{S}(k\rho_H)$ (rather than $\Sigma^{-1}\widehat{S}(k\rho_H)$) are said to be* pure.

In order to define $\mathcal{S}^G_{>m}$, we need to assign a dimension to each element in $\mathcal{A}$. We do this in terms of the underlying wedge summands, namely

$$\dim \widehat{S}(k\rho_H) = kh \quad \text{and} \quad \dim \Sigma^{-1}\widehat{S}(k\rho_H) = kh - 1.$$

Then $\mathcal{S}^G_{>m}$ is the category built up out of elements in $\mathcal{A}$ of dimension $> m$ using arbitrary wedges, mapping cones and smash products with equivariant suspension spectra.

3.3. The slice tower

With this definition it is possible to construct functors P^G_{m+1} and P^m_G with the same formal properties as in the classical case. Thus we get a tower

$$\begin{array}{ccccccccc} \cdots \longrightarrow & P_G^{m+1}X & \longrightarrow & P_G^m X & \longrightarrow & P_G^{m-1}X & \longrightarrow \cdots \\ & \uparrow & & \uparrow & & \uparrow & \\ & {}^GP^{m+1}_{m+1}X & & {}^GP^m_m X & & {}^GP^{m-1}_{m-1}X & \end{array} \qquad (2)$$

in which the homotopy colimit is X and the homotopy limit is contractible.

We call this the *slice tower.* ${}^GP^m_m X$ is the *nth slice* and the decreasing sequence of subgroups of $\pi_* X$ is the *slice filtration.* We also get slice filtrations of the $RO(G)$-graded homotopy $\pi^G_\star X$ and the homotopy groups of fixed point sets $\pi_* X^H$. *From now on we will drop the symbol G from the functors P^m, P_m and P^m_m.*

There is an important difference between this tower and the classical one. In the classical case the map $X \to P^m X$ does not change homotopy groups in dimensions $\leq m$. *This is not true in this equivariant case.*

In the classical case, $P_m^m X$ is an Eilenberg-Mac Lane spectrum whose nth homotopy group is that of X. In our case, $\pi_* P_m^m X$ need not be concentrated in dimension m.

This means the slice filtration leads to a (possibly noncollapsing) *slice spectral sequence* converging to $\pi_* X$ and its variants.

One variant has the form

$$E_2^{s,t} = \pi_{t-s}^G P_t^t X \implies \pi_{t-s}^G X.$$

Recall that $\pi_*^G X$ is by definition $\pi_* X^G$, the homotopy of the fixed point set.

This is the spectral sequence we will use to study $MU^{(G)}$ and its relatives.

3.4. The Slice Theorem

A large portion of our paper is devoted to proving that the slice spectral sequence has the desired properties.

Slice Theorem. *In the slice tower for $MU^{(G)}$, every odd slice is contractible and P_{2m}^{2m} decomposes as $\widehat{W}_{2m} \wedge H\mathbf{Z}_{(2)}$, where $\widehat{W}_{2m}$ is a certain wedge of $2m$-dimensional slice cells (to be named later) and $H\mathbf{Z}_{(2)}$ is the 2-local integer Eilenberg-Mac Lane spectrum with trivial G action. $\widehat{W}_{2m}$ never has any free summands.*

We will encounter other spectra with similar properties, so we give them a name.

Definition 3.2. *A 2-local spectrum X is* pure (isotropic) *if each of its odd slices is contractible and each even slice has the form $\widehat{W} \wedge H\mathbf{Z}_{(2)}$ where $\widehat{W}$ is a wedge of pure (isotropic) cells (see Definition 3.1).*

In order to specify $\widehat{W}_{2m}$, we need the following definition.

Definition 3.3. *Suppose X is a (2-local) G-spectrum such that its underlying homotopy group $\pi_k^u X$ is a free abelian group (a free $\mathbf{Z}_{(2)}$-module). A* refinement of $\pi_k^u X$ *is an equivariant map*

$$c : \widehat{W} \to X$$

in which $\widehat{W}$ is a wedge of slice cells of dimension k whose underlying spheres represent a basis of $\pi_k^u X$.

Recall that in $\pi_*^u(MU_{\mathbf{R}})$, any monomial in the polynomial generators in dimension $2m$ is represented by an equivariant map from $S^{m\rho_2}$.

$\pi_*^u MU^{(C_8)}$ is a polynomial algebra with 4 generators in every positive even dimension. We will denote the generators in dimension $2i$ by $\gamma^j r_i$ for $0 \leq j \leq 3$. The action of $\gamma^4 \in G = C_8$ is given by

$$\gamma^4 r_i = (-1)^i r_i.$$

We will explain how $\pi_*^u MU^{(C_8)}$ can be refined.

$\pi_2^u MU^{(C_8)}$ has 4 generators $\gamma^j r_1$ that are permuted up to sign by G. It is refined by an equivariant map

$$\widehat{W}_2 = \widehat{S}(\rho_2) \to MU^{(C_8)}.$$

Recall that the underlying spectrum of $\widehat{S}(\rho_2)$ is a wedge of 4 copies of S^2.

In $\pi_4^u MU^{(C_8)}$ there are 14 monomials that fall into 4 orbits under the action of G, each corresponding to a map from a slice cell.

$$\begin{array}{rcl}
\widehat{S}(2\rho_2) & \longleftrightarrow & \{r_1^2,\ \gamma r_1^2,\ \gamma^2 r_1^2,\ \gamma^3 r_1^2\} \\
\widehat{S}(2\rho_2) & \longleftrightarrow & \{r_1 \cdot \gamma r_1,\ \gamma r_1 \cdot \gamma^2 r_1,\ \gamma^2 r_1 \cdot \gamma^3 r_1,\ \gamma^3 r_1 \cdot r_1\} \\
\widehat{S}(2\rho_2) & \longleftrightarrow & \{r_2,\ \gamma r_2,\ \gamma^2 r_2,\ \gamma^3 r_2\} \\
\widehat{S}(\rho_4) & \longleftrightarrow & \{r_1 \cdot \gamma^2 r_1,\ \gamma r_1 \cdot \gamma^3 r_1\}
\end{array}$$

Recall that $\widehat{S}(\rho_4)$ is underlain by $S^4 \vee S^4$. It follows that $\pi_4^u MU^{(C_8)}$ is refined by an equivariant map from

$$\widehat{W}_4 = \widehat{S}(2\rho_2) \vee \widehat{S}(2\rho_2) \vee \widehat{S}(\rho_4) \vee \widehat{S}(2\rho_2).$$

3.5. Refining $\pi_*^u MU^{(G)}$

More generally, we can proceed as follows. If we regard $MU^{(G)}$ as a C_2-spectrum, then $\pi_*^u MU^{(G)}$ is refined by a map from a wedge of $S^{k\rho_2}$s. More specifically, for each $i > 0$ we have a map

$$S^0[\overline{r}_i] = \bigvee_{m \geq 0} S^{mi\rho_2} \to MU^{(G)}$$

representing the powers of a certain generator $\overline{r}_i \in \pi_{2i}^{C_2} MU^{(G)}$ defined in [HHR, §5.4.2]. Using the multiplication on $MU^{(G)}$, we get a map

$$S^0[\overline{r}_1, \overline{r}_2, \ldots] = \bigwedge_{i>0} S^0[\overline{r}_i] \to MU^{(G)}.$$

This spectrum is a wedge of $S^{m\rho_2}$s representing each monomial in the r_is.

Now we apply the internal norm functor N_2^g to this map and get a map

$$A = N_2^g S^0[\overline{r}_1, \overline{r}_2, \ldots] \to MU^{(G)} \tag{3}$$

which refines the underlying homotopy of the target as a G-spectrum. A is a wedge of even dimensional slice cells, and the $\widehat{W}_{2m}$ in the Slice Theorem is the wedge of $2m$-dimensional slice cells in A.

3.6. The definition of $\Omega_{\mathbb{O}}$

Our G-spectrum $\Omega_{\mathbb{O}}$ (where $G = C_8$) is obtained from the E_∞-ring spectrum $MU^{(G)}$ by inverting a certain element $D \in \pi_{19\rho_G}^G$ specified in (6) below. The choices of G and D are the simplest ones leading to a homotopy fixed point set with the detection property, as shown in [HHR, §11]. The slice tower for $\Omega_{\mathbb{O}}$ has similar properties to that of $MU^{(G)}$.

4. The Gap Theorem

4.1. Derivation from the Slice Theorem

Assuming the Slice Theorem, the Gap Theorem follows from the fact that π_{-2}^G vanishes *for each isotropic slice*, i.e., for each one of the form

$$\widehat{S}(k\rho_H) \wedge H\mathbf{Z}_{(2)}$$

for nontrivial H. Proving this amounts to computing the homology of a certain chain complex. Details can be found in [HHR, §3.2]. *These calculations do not require localization at 2, so we use $H\mathbf{Z}$ instead of $H\mathbf{Z}_{(2)}$ thoughout this section.*

In order to give a feel for these calculations, a picture of $\pi_*^G S^{k\rho_G} \wedge H\mathbf{Z}$ for $G = C_8$ and various integers k is shown in Figure 1.

In order to derive the Gap Theorem from the Slice Theorem we need to find the groups

$$\pi_*^G W(m\rho_h) \wedge H\mathbf{Z} = \pi_*^H S^{m\rho_h} \wedge H\mathbf{Z}.$$

We need this for *all* integers m because eventually we will obtain the spectrum $\Omega_{\mathbb{O}}$ by inverting a certain element in $\pi_{19\rho_8}^G MU^{(C_8)}$. Here is what we will learn.

Vanishing Theorem.

- *For $m \geq 0$, $\pi_k^H S^{m\rho_h} \wedge H\mathbf{Z} = 0$ unless $m \leq k \leq mh$.*
- *For $m < 0$ and $h > 1$, $\pi_k^H S^{m\rho_h} \wedge H\mathbf{Z} = 0$ unless $mh \leq k \leq m-2$. The upper bound can be improved to $m-3$ except in the case $(h, m) = (2, -2)$ in which case we have $\pi_{-4}^H S^{-2\rho_2} \wedge H\mathbf{Z} = \mathbf{Z}_{(2)}$.*

Gap Corollary. *For $h > 1$ and all integers m, $\pi_k^H S^{m\rho_h} \wedge H\mathbf{Z} = 0$ for $-4 < k < 0$.*

This means a similar statement must hold for $\pi_*^{C_8}\Omega_{\mathbb{O}} = \pi_*\Omega$, which gives the Gap Theorem.

Assuming the Slice Theorem, the Gap Theorem (the statement that $\pi_{-2}\Omega = 0$) follows immediately from the Gap Corollary above, as does the following.

Vanishing Line Corollary. *The slice spectral sequence for $\pi_*^G MU^{(G)}$ is confined to the first quadrant and*

$$E_2^{s,t} = 0 \qquad \text{for } s > (g-1)(t-s),$$

where $g = |G|$. If we invert any element in $\pi_{k\pi_G}^G MU^{(G)}$, the spectral sequence is confined to the first and third quadrants, and in the latter

$$E_2^{s,t} = 0 \qquad \text{for } s < \min(0, (g-1)(t-s+4)+4-g).$$

The proofs of the Vanishing Theorem and Gap Corollary are surprisingly easy.

4.2. An easy calculation

We begin by constructing an equivariant cellular chain complex $C(m\rho_g)_*$ for $S^{m\rho_g}$, where $m \geq 0$. In it the cells are permuted by the action of G, and the action of any

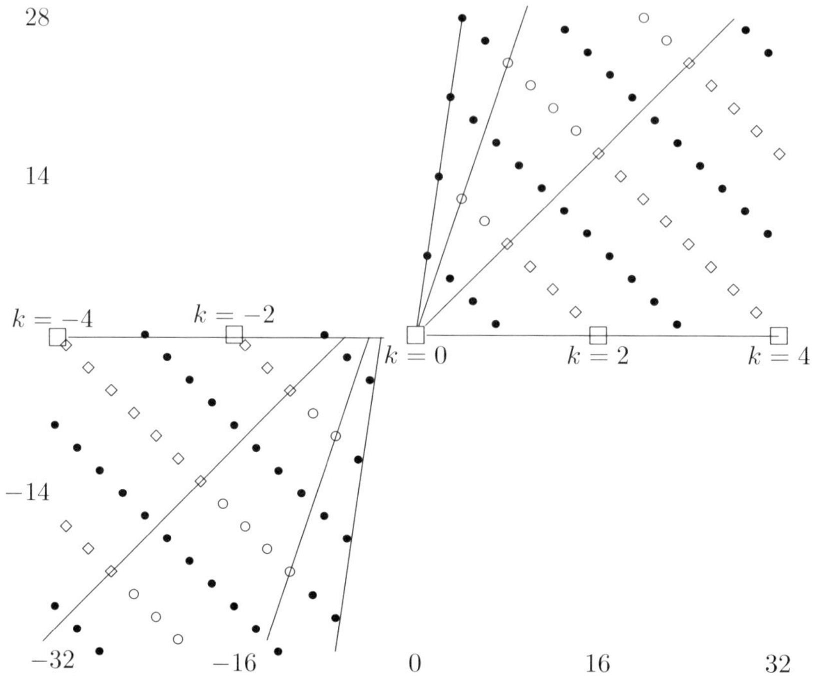

FIGURE 1. The homotopy of the slices $S^{k\rho_8} \wedge H\mathbf{Z}$ for small k. The horizontal and vertical axes are $t-s$ and s in the slice spectral sequence. The numbers along the $(t-s)$-axis are even values of k. The groups $\pi_*^G S^{k\rho_8} \wedge H\mathbf{Z}$ lie along diagonals where $t = 8k$. Squares, bullets, circles and diamonds stand for $\mathbf{Z}$, $\mathbf{Z}/2$, $\mathbf{Z}/4$ and $\mathbf{Z}/8$ respectively. Lines of slopes 1, 3 and 7 separate regions with various types of torsion for even k. *The gap can be seen in the center of this chart.* The vanishing regions seen here also occur in the slice spectral sequence for $\pi_*^G MU^{(G)}$.

subgroup H on the H-invariant cells is trivial. It is a complex of $\mathbf{Z}[G]$-modules and is uniquely determined by fixed point data of $S^{m\rho_g}$.

For $H \subset G$ we have

$$(S^{m\rho_g})^H = S^{mg/h}.$$

This means there is a G-CW-complex with one cell in dimension m, two cells in each dimension from $m+1$ to $2m$, four cells in each dimension from $2m+1$ to $4m$, and so on.

In other words,

$$C(m\rho_g)_k = \begin{cases} 0 & \text{unless } m \le k \le gm \\ \mathbf{Z} & \text{for } k = m \\ \mathbf{Z}[G/H] & \text{for } mg/2h < k \le mg/h \text{ and } h < g. \end{cases}$$

Each of these is a cyclic $\mathbf{Z}[G]$-module. The boundary operator is uniquely determined by the fact that $H_*(C(m\rho_g)) = H_*(S^{gm})$.

Then we have

$$\pi_*^G(S^{m\rho_g} \wedge H\mathbf{Z}) = H_*(\mathrm{Hom}_{\mathbf{Z}[G]}(\mathbf{Z}, C(m\rho_g))).$$

These groups are nontrivial only for $m \le k \le gm$, which gives the Vanishing Theorem for $m \ge 0$.

We will look at the bottom three groups in the complex $\mathrm{Hom}_{\mathbf{Z}[G]}(\mathbf{Z}, C(m\rho_g)_*)$. Since $C(m\rho_g)_k$ is a cyclic $\mathbf{Z}[G]$-module, the Hom group is always $\mathbf{Z}$.

For $m > 1$ our chain complex $C(m\rho_g)$ has the form

$$\begin{array}{ccccccc} & & C(m\rho_g)_m & & C(m\rho_g)_{m+1} & & C(m\rho_g)_{m+2} \\ & & \| & & \| & & \| \\ 0 & \longleftarrow & \mathbf{Z} & \xleftarrow{\epsilon} & \mathbf{Z}[C_2] & \xleftarrow{1-\gamma} & \mathbf{Z}[C_2] & \xleftarrow{1+\gamma} \cdots \end{array}$$

Applying $\mathrm{Hom}_{\mathbf{Z}[G]}(\mathbf{Z}, \cdot)$ to this gives (in dimensions $\le 2m$ for $m > 4$)

$$\begin{array}{ccccccccc} \mathbf{Z} & \xleftarrow{2} & \mathbf{Z} & \xleftarrow{0} & \mathbf{Z} & \xleftarrow{2} & \mathbf{Z} & \xleftarrow{0} & \mathbf{Z} \longleftarrow \cdots \\ m & & m+1 & & m+2 & & m+3 & & m+4 \end{array}$$

It follows that for $m \le k < 2m$,

$$\pi_k^G(S^{m\rho_g} \wedge H\mathbf{Z}) = \begin{cases} \mathbf{Z}/2 & k \equiv m \bmod 2 \\ 0 & \text{otherwise.} \end{cases}$$

For negative multiples of ρ_g, $S^{-m\rho_g}$ (with $m > 0$) is the equivariant Spanier-Whitehead dual of $S^{m\rho_g}$. This means that

$$\pi_*^G S^{-m\rho_g} \wedge H\mathbf{Z} = H^*(\mathrm{Hom}_{\mathbf{Z}[G]}(C(m\rho_g), \mathbf{Z})).$$

Applying the functor $\mathrm{Hom}_{\mathbf{Z}[G]}(\cdot, \mathbf{Z})$ to our chain complex $C(m\rho_g)$

$$\begin{array}{ccccccc} \mathbf{Z} & \xleftarrow{\epsilon} & \mathbf{Z}[C_2] & \xleftarrow{1-\gamma} & \mathbf{Z}[C_2] & \xleftarrow{1+\gamma} & \mathbf{Z}[C_2 \text{ or } C_4] \xleftarrow{1-\gamma} \cdots \\ m & & m+1 & & m+2 & & m+3 \end{array}$$

gives a cochain complex beginning with

$$\begin{array}{ccccccccc} \mathbf{Z} & \xrightarrow{1} & \mathbf{Z} & \xrightarrow{0} & \mathbf{Z} & \xrightarrow{2} & \mathbf{Z} & \xrightarrow{0} & \mathbf{Z} \longrightarrow \cdots \\ -m & & -m-1 & & -m-2 & & -m-3 & & -m-4 \end{array}$$

Here is a diagram showing both functors in the case $m \geq 4$.

$$
\begin{array}{ccccccccccc}
m & & m+1 & & m+2 & & m+3 & & m+4 & & \\
\mathbf{Z} & \xleftarrow[2]{} & \mathbf{Z} & \xleftarrow[0]{} & \mathbf{Z} & \xleftarrow[2]{} & \mathbf{Z} & \xleftarrow[0]{} & \mathbf{Z} & \longleftarrow & \ldots \\
 & & & & \Big\Uparrow \mathrm{Hom}_{\mathbf{Z}[G]}(\mathbf{Z}, \cdot) & & & & & & \\
\mathbf{Z} & \xleftarrow{\epsilon} & \mathbf{Z}[C_2] & \xleftarrow{1-\gamma} & \mathbf{Z}[C_2] & \xleftarrow{1+\gamma} & \mathbf{Z}[C_2] & \xleftarrow{1-\gamma} & \mathbf{Z}[C_2] & \xleftarrow{1-\gamma} & \ldots \\
 & & & & \Big\Downarrow \mathrm{Hom}_{\mathbf{Z}[G]}(\cdot, \mathbf{Z}) & & & & & & \\
\mathbf{Z} & \xrightarrow{1} & \mathbf{Z} & \xrightarrow{0} & \mathbf{Z} & \xrightarrow{2} & \mathbf{Z} & \xrightarrow{0} & \mathbf{Z} & \longrightarrow & \ldots \\
-m & & -m-1 & & -m-2 & & -m-3 & & -m-4 & &
\end{array}
$$

Note the difference in the behavior of the map $\epsilon : \mathbf{Z}[C_2] \to \mathbf{Z}$ *under the two functors* $\mathrm{Hom}_{\mathbf{Z}[G]}(\mathbf{Z}, \cdot)$ *and* $\mathrm{Hom}_{\mathbf{Z}[G]}(\cdot, \mathbf{Z})$*. They convert it to maps of degrees 2 and 1 respectively. This difference is responsible for the gap.*

5. The Periodicity Theorem

We now outline the proof of the Periodicity Theorem (assuming the Slice Theorem), which is treated in [HHR, §9].

We establish some differentials in the slice spectral sequence and show that certain elements become permanent cycles after inverting a certain $D \in \pi^G_{19\rho_G} MU^{(G)}$ (defined below in (6)), where $G = C_8$. This leads to an equivariant self map

$$\Sigma^{256}\Omega_{\mathbb{O}} \to \Omega_{\mathbb{O}}.$$

It is an ordinary homotopy equivalence, and we will see that this implies formally that it induces an equivalence on homotopy fixed point sets.

5.1. Geometric fixed points

The key tool for studying differentials in the slice spectral sequence is the *geometric fixed point spectrum*, denoted by $\Phi^G X$ for a G-spectrum X. A detailed account and references can be found in [HHR, §2.5]. It has much nicer properties than the usual fixed point spectrum X^G, which is awkward for two reasons:

- it fails to commute with smash products and
- it fails to commute with infinite suspensions.

The *geometric fixed set* $\Phi^G X$ is a convenient substitute that avoids these difficulties. In order to define it we need the *isotropy separation sequence*, which is the cofiber sequence

$$E\mathcal{P}_+ \to S^0 \to \tilde{E}\mathcal{P}. \tag{4}$$

where $E\mathcal{P}$ is a G-space such that for a subgroup H of G,

$$E\mathcal{P}^H = \begin{cases} \text{empty} & \text{for } H = G \\ \text{contractible} & \text{otherwise.} \end{cases}$$

Such a space can be constructed for any finite group G and it is unique up to equivariant homotopy equivalence. For cyclic 2-group G, the space EC_2 with G acting via it projection onto C_2 has the required properties. The symbol $\mathcal{P}$ here stands for the family of proper subgroups of G. A similar space $E\mathcal{F}$ can be constructued for any family of subgroups $\mathcal{F}$ closed under inclusion and conjugation.

Definition 5.1. *For a G-spectrum X the* geometric fixed point set *is*

$$\Phi^G X = (\tilde{E\mathcal{P}} \wedge X)^G.$$

This functor has the following properties:

Theorem 5.2.

(i) *For G-spectra X and Y, $\Phi^G(X \wedge Y) = \Phi^G X \wedge \Phi^G Y$.*
(ii) *For a G-space X, $\Phi^G \Sigma^\infty X = \Sigma^\infty(X^G)$.*
(iii) *A map $f : X \to Y$ is a G-equivalence iff $\Phi^H f$ is an ordinary equivalence for each subgroup $H \subset G$.*

From the suspension property we can deduce that for any finite cyclic 2-group G,

$$\Phi^G MU^{(G)} = MO,$$

the unoriented cobordism spectrum. Its homotopy type has been well understood since Thom's work [Tho54] in the 50s.

Geometric Fixed Point Theorem. *Let G be a finite cyclic 2-group and let σ denote its sign representation. Then for any G-spectrum X, $\pi_\star \tilde{E}C_2 \wedge X = a_\sigma^{-1}\pi_\star X$, where $a_\sigma \in \pi_{-\sigma} X$ is induced by the inclusion of the fixed point set $S^0 \to S^\sigma$.*

In [HHR, §5.4.2] we define specific polynomial generators

$$\overline{r}_i^G \in \pi_{i\rho_2}^{C_2} MU^{(G)} \tag{5}$$

that are convenient for our purposes. We denote the underlying homotopy classes by r_i that lie in $\pi_{2i}^u MU^{(G)}$.

Recall that $\pi_* MO = \mathbf{Z}/2[y_i : i > 0, i \neq 2^k - 1]$ where $|y_i| = i$. It is not hard to show that

$$\pi_*^u MU^{(C_8)} = \mathbf{Z}_{(2)}[r_i, \gamma(r_i), \gamma^2(r_i), \gamma^3(r_i) : i > 0]$$

where $|r_i| = 2i$, γ is a generator of G and $\gamma^4(r_i) = (-1)^i r_i$. In $\pi_{i\rho_8} MU^{(4)}$ we have the element

$$N r_i = r_i \gamma(r_i) \gamma^2(r_i) \gamma^3(r_i).$$

Applying the functor Φ^G to the map $N_2^8 r_i : S^{i\rho_8} \to MU^{(C_8)}$ gives a map $S^i \to MO$.

Lemma 5.3. *The generators r_i and y_i satisfy*

$$\Phi^G N_2^g r_i = \begin{cases} 0 & \textit{for } i = 2^k - 1 \\ y_i & \textit{otherwise.} \end{cases}$$

5.2. Some slice differentials

We know that the slice spectral sequence for $MU^{(G)}$ has a vanishing line of slope $g-1$. We will describe the subring of elements lying on it.

Let $f_i \in \pi_i(MU^{(G)})$ be the composite

$$S^i \xrightarrow{a_{i\rho_g}} S^{i\rho_g} \xrightarrow{Nr_i} MU^{(G)},$$

where $a_{i\rho_g}$ is the inclusion of the fixed point set. The following facts about f_i are easy to prove.

- It appears in the slice spectral sequence in $E_2^{(g-1)i,gi}$, which is on the vanishing line.
- The subring of elements on the vanishing line is the polynomial algebra on the f_i.
- Under the map

  $$\pi_* MU^{(G)} \to \pi_* \Phi^G MU^{(G)} = \pi_* MO$$

  we have

  $$f_i \mapsto \begin{cases} 0 & \text{for } i = 2^k - 1 \\ y_i & \text{otherwise} \end{cases}$$

- Any differential landing on the vanishing line must have a target in the ideal given by
  $(f_1, f_3, f_7, \ldots)$. A similar statement can be made after smashing with $S^{2^k\sigma}$.

For an oriented representation V there is a map $u_V : S^{|V|} \to \Sigma^V H\mathbf{Z}_{(2)}$, which lies in $\pi_{V-|V|}(H\mathbf{Z}_{(2)})$. It satisfies $u_{V+W} = u_V u_W$, so $u_{2^k\sigma} = u_{2\sigma}^{2^{k-1}}$.

Slice Differentials Theorem. *In the slice spectral sequence for $\Sigma^{2^k\sigma} MU^{(G)}$ for $k > 0$, we have $d_r(u_{2^k\sigma}) = 0$ for $r < 1 + g(2^k - 1)$, and*

$$d_{1+g(2^k-1)}(u_{2^k\sigma}) = a_\sigma^{2^k} f_{2^k-1}.$$

Sketch of proof: Inverting a_σ in the slice spectral sequence will make it converge to $\pi_*(MO)$. This means each power of $u_{2\sigma}$ has to support a nontrivial differential. The only way this can happen is as indicated in the theorem.

Typically one proves theorems about differentials in such spectral sequences by means of some sort of extended power construction. In our case, all of the necessary geometry is encoded in the relation between $\pi_*^u MU^{(G)}$ and $\pi_* MO$!

5.3. Some $RO(G)$-graded calculations

For a cyclic 2-group G let

$$\begin{aligned}\overline{\mathfrak{d}}_k^G = N_2^g \overline{r}_{2^k-1} &= \overline{r}_{2^k-1}\gamma(\overline{r}_{2^k-1})\dots\gamma^{g/2-1}(\overline{r}_{2^k-1}) \\ &\in \pi_{(2^k-1)\rho_g} MU^{(G)}\end{aligned}$$

We want to invert this element and study the resulting slice spectral sequence. As explained previously, for $G = C_8$ it is confined to the first and third quadrants with vanishing lines of slopes 0 and 7.

The differential d_r on $u_{2\sigma}^{2^k}$ described in the theorem is the last one possible since its target, $a_\sigma^{2^{k+1}} f_{2^{k+1}-1}$, lies on the vanishing line. If we can show that this target is killed by an earlier differential after inverting $\overline{\mathfrak{d}}_k^G$, then $u_{2\sigma}^{2^k}$ will be a permanent cycle.

We have

$$\begin{aligned}f_{2^{k+1}-1}\overline{\mathfrak{d}}_k^G &= (a_{\rho_g}^{2^{k+1}-1} N_2^g \overline{r}_{2^{k+1}-1})(N_2^g \overline{r}_{2^k-1}) \\ &= a_{\rho_g}^{2^k} N_2^g \overline{r}_{2^{k+1}-1}(a_{\rho_g}^{2^k-1} N_2^g \overline{r}_{2^k-1}) \\ &= a_{\rho_g}^{2^k} \overline{\mathfrak{d}}_{k+1}^G f_{2^k-1} \\ &= a_V^{2^k} \overline{\mathfrak{d}}_{k+1}^G a_\sigma^{2^k} f_{2^k-1} \qquad \text{where } V = \rho_g - \sigma \\ &= a_V^{2^k} p\overline{\mathfrak{d}}_{k+1}^G d_{1+8(2^k-1)}(u_{2^k\sigma}).\end{aligned}$$

Corollary. *In the $RO(G)$-graded slice spectral sequence for $\left(\overline{\mathfrak{d}}_k^G\right)^{-1} MU^{(G)}$, the class given by $u_{2^{k+1}\sigma} = u_{2\sigma}^{2^k}$ is a permanent cycle.*

5.4. An even trickier $RO(G)$-graded calculation

The Corollary shows that inverting a certain element makes a power of $u_{2\sigma}$ a permanent cycle. *We need to invert something to make a power of $u_{2\rho_8}$ a permanent cycle.*

We will get this by using the norm property of u. It says that if V is an oriented representation of a subgroup $H \subset G$ with $V^H = 0$ and V' is the induced representation of V, then the norm functor N_h^g from H-spectra to G-spectra satisfies $N_h^g(u_V)u_{V''} = u_{V'}$, where V'' is the induced representation of the trivial representation of degree $|V|$.

From this we can deduce that

$$u_{2\rho_8} = u_{8\sigma_8} N_4^8(u_{4\sigma_4}) N_2^8(u_{2\sigma_2}),$$

where σ_g denotes the sign representation on C_g.

By the Corollary we can make a power of each factor a permanent cycle by inverting some $\overline{\mathfrak{d}}_{k_m}^{C_{2^m}}$ for $1 \le m \le 3$. If we make k_m too small we will lose the detection property, that is we will get a spectrum that does not detect the θ_j. It turns out that k_m must be chosen so that $8|2^m k_m$.

- Inverting $\overline{\mathfrak{d}}_4^{C_2}$ makes $u_{32\sigma_2}$ a permanent cycle.

- Inverting $\bar{\mathfrak{d}}_2^{C_4}$ makes $u_{8\sigma_4}$ a permanent cycle.
- Inverting $\bar{\mathfrak{d}}_1^{C_8}$ makes $u_{4\sigma_8}$ a permanent cycle.
- Inverting the product D of the norms of all three makes $u_{32\rho_8} = u_{2\rho_8}^{16}$ a permanent cycle.

Let

$$D = \bar{\mathfrak{d}}_1^{C_8} N_4^8(\bar{\mathfrak{d}}_2^{C_4}) N_2^8(\bar{\mathfrak{d}}_4^{C_2}) = N_2^8(\bar{r}_1^{C_8}\bar{r}_3^{C_4}\bar{r}_{15}^{C_2}) \in \pi_{19\rho_8} MU^{(C_8)}. \tag{6}$$

Then we define

$$\Omega_{\mathbb{O}} = D^{-1} MU^{(C_8)} \qquad \text{and} \qquad \Omega = \Omega_{\mathbb{O}}^{hC_8}. \tag{7}$$

Since the inverted element is represented by a map from $S^{m\rho_8}$, the slice spectral sequence for $\pi_*\Omega = \pi_*^{C_8}\Omega_{\mathbb{O}}$ has the usual properties:

- It is concentrated in the first and third quadrants and confined by vanishing lines of slopes 0 and 7.
- It has the gap property, i.e., no homotopy between dimensions -4 and 0.

5.5. The proof of the Periodicity Theorem

Preperiodicity Theorem. *Let* $\Delta_1^{(8)} = u_{2\rho_8}\left(\bar{\mathfrak{d}}_1^{C_8}\right)^2 \in E_2^{16,0}(\Omega_{\mathbb{O}})$. *Then* $\left(\Delta_1^{(8)}\right)^{16}$ *is a permanent cycle.*

To prove this, note that $\left(\Delta_1^{(8)}\right)^{16} = u_{32\rho_8}\left(\bar{\mathfrak{d}}_1^{C_8}\right)^{32}$. Both $u_{32\rho_8}$ and $\bar{\mathfrak{d}}_1^{C_8}$ are permanent cycles, so $\left(\Delta_1^{(8)}\right)^{16}$ is also one.

Hence we have an equivariant map $\Pi : \Sigma^{256}\Omega_{\mathbb{O}} \to \Omega_{\mathbb{O}}$ where

- $u_{32\rho_8} : S^{256-32\rho_8} \to \Omega_{\mathbb{O}}$ induces to the unit map from S^0 on the underlying ring spectrum and
- $\Delta_1^{(8)}$ is invertible because it is a factor of D.

The above imply that the underlying map $i_0\Pi$ of ordinary spectra is a homotopy equivalence. It is known that any such map induces an equivalence of homotopy fixed point sets, so

$$\Sigma^{256}\Omega_{\mathbb{O}}^{hC_8} \xrightarrow[\simeq]{\Pi^{hC_8}} \Omega_{\mathbb{O}}{}^{hC_8}.$$

Unfortunately the slice spectral sequence tells us nothing about this homotopy fixed point set. We know it detects all of the θ_j, but there is no direct way of showing that it has the gap property.

Fortunately we have a theorem (the subject of the next section) stating that in this case the homotopy fixed set is equivalent to the actual fixed point set Ω. The slice spectral sequence tells us that the latter has the gap property. Thus we have proved

Periodicity Theorem. *For* Ω *as in (7),* $\Sigma^{256}\Omega$ *is equivalent to* Ω.

6. The Homotopy Fixed Point Theorem

In order to finish the proof of the main theorem, we need to show that the actual fixed point set of $\Omega_{\mathbb{O}} = D^{-1}MU^{(C_8)}$ is equivalent to its homotopy fixed point set. A detailed account can be found in [HHR, §10]. The slice spectral sequence computes the homotopy of the former while the Hopkins-Miller spectral sequence, also known as the homotopy fixed point spectral sequence (which is known to detect θ_j), computes that of the latter.

Here is a general approach to showing that actual and homotopy fixed points are equivalent for a G-spectrum X.

We have an equivariant map $EG_+ \to S^0$. Mapping both into X gives a map of G-spectra $\varphi : X_+ \to F(EG_+, X_+)$. Passing to fixed points would give a map from X^G to X^{hG}, which we would like to be an equivalence. We will prove the stronger statement that φ is a G-equivalence.

The case of interest is $X = \Omega_{\mathbb{O}}$ and $G = C_8$. We will argue by induction on the order of the subgroups H of G, the statement being obvious for the trivial group. We will smash φ with the cofiber sequence

$$EG_+ \to S^0 \to \tilde{E}G.$$

This gives us the following diagram in which both rows are cofiber sequences.

$$\begin{array}{ccccc}
EG_+ \wedge \Omega_{\mathbb{O}} & \longrightarrow & \Omega_{\mathbb{O}} & \longrightarrow & \tilde{E}G \wedge \Omega_{\mathbb{O}} \\
\downarrow{\scriptstyle \varphi'} & & \downarrow{\scriptstyle \varphi} & & \downarrow{\scriptstyle \varphi''} \\
EG_+ \wedge F(EG_+, \Omega_{\mathbb{O}}) & \longrightarrow & F(EG_+, \Omega_{\mathbb{O}}) & \longrightarrow & \tilde{E}G \wedge F(EG_+, \Omega_{\mathbb{O}})
\end{array}$$

The map φ' is an equivalence because $\Omega_{\mathbb{O}}$ is nonequivariantly equivalent to $F(EG_+, \Omega_{\mathbb{O}})$, and EG_+ is built up entirely of free G-cells.

Thus it suffices to show that φ'' is an equivalence, which we will do by showing that both its source and target are contractible. Both have the form $\tilde{E}G \wedge X$ where X is a module spectrum over $\Omega_{\mathbb{O}}$, so it suffices to show that $\tilde{E}G \wedge \Omega_{\mathbb{O}}$ is contractible.

We will show that it is H-equivariantly contractible by induction on the order of the subgroups H of G. Over the trivial group $\tilde{E}G$ itself is contractible. Let H be a subgroup, $H' \subset H$ the subgroup of index 2 and $H_2 = H/H'$.

We will smash our spectrum with the isotropy separation sequence of (4), which in our case is

$$EH_{2+} \to S^0 \to \tilde{E}H_2.$$

Then $\tilde{E}H_2 \wedge \tilde{E}G \wedge \Omega_{\mathbb{O}}$ is contractible over H', so it suffices to show that its H-fixed point set is contractible. It is

$$\Phi^H(\tilde{E}G \wedge \Omega_{\mathbb{O}}) = \Phi^H(\tilde{E}G) \wedge \Phi^H(\Omega_{\mathbb{O}}),$$

and $\Phi^H(\Omega_{\mathbb{O}})$ is contractible because $\Phi^H(D) = 0$. Thus it remains to show the H-contractibility of $EH_{2+} \wedge \tilde{E}G \wedge \Omega_{\mathbb{O}}$. But this is equivalent to the H'-contractibility of $\tilde{E}G \wedge \Omega_{\mathbb{O}}$, which we have by induction.

Sidebar 3 Formal group laws I

Formal groups laws are are key tool in complex cobordism theory, which is part of the infrastructure of our proof. The definitive reference is Hazewinkel's book [Haz78]. A much briefer account can be found in [Rav86, Appendix 2]. A *formal group law F over a ring R* is a power series $F(x, y) \in R[[x, y]]$ satisfying three conditions, namely

(i) COMMUTATIVITY $F(y, x) = F(x, y)$
(ii) IDENTITY $F(x, 0) = F(0, x) = x$
(iii) ASSOCIATIVITY $F(F(x, y), z) = F(x, F(y, z))$.

Elementary examples include the additive and multiplicative formal group laws, $x + y$ and $1 + x + y + xy$ respectively. When R is torsion free, F is determined by a power series over $R \otimes \mathbf{Q}$ in one variable called the *logarithm of F* satisfying

$$\log_F(F(x, y)) = \log_F(x) + \log_F(y).$$

For our two elementary examples it is x and $\sum_{n \geq 0} (-1)^n x^{n+1}/(n + 1)$.

There is a universal example due to Lazard [Laz55] of a formal group law G over a ring L with the property that for any formal group law F over any ring R, there is a unique ring homomorphism $\theta : L \to R$ sending G to F. It is given by

$$G(x, y) = \sum_{i,j} a_{i,j} x^i y^j \quad \text{and} \quad L = \mathbf{Z}[a_{i,j}]/I$$

where I is the ideal generated by the relations among the coefficients $a_{i,j}$ forced by conditions (i)-(iii) above. To describe L explicitly, it is useful to give it a grading such that $G(x, y)$ is homogeneous of degree 2 if $|x| = |y| = 2$. This means that $|a_{i,j}| = 2(1 - i - j)$. Lazard then shows that

$$L = \mathbf{Z}[x_i : i > 0] \qquad \text{where } |x_i| = -2i.$$

7. The Detection Theorem

The Detection Theorem is the subject of [HHR, §11].

7.1. θ_j in the Adams-Novikov spectral sequence

Browder's Theorem says that θ_j is detected in the classical Adams spectral sequence by

$$h_j^2 \in \operatorname{Ext}_{\mathscr{A}}^{2,2^{j+1}}(\mathbf{Z}/2, \mathbf{Z}/2),$$

where $\mathscr{A}$ denotes the mod 2 Steenrod algebra. This element is known to be the only one in its bidegree.

Sidebar 4 Formal group laws II

There is a formal group law F defined over the complex cobordism ring MU^* defined as follows. We know that
$$MU^*(\mathbf{C}P^\infty) = MU^*[[x]]$$
where $x \in MU^2\mathbf{C}P^\infty$ is the first Chern class of the canonical complex line bundle. Similarly,
$$MU^*(\mathbf{C}P^\infty \times \mathbf{C}P^\infty) = MU^*[[x \otimes 1, 1 \otimes x]].$$
There is a map $\mathbf{C}P^\infty \times \mathbf{C}P^\infty \to \mathbf{C}P^\infty$ related to the tensor product of line bundles. It sends x to a power series $F(x \otimes 1, 1 \otimes x)$, which can easily be shown to be a formal group law. Quillen [Qui69] showed that the Lazard classifing map $L \to MU^*$ is an isomorphism.

Cartier [Car67] showed that when R is a torsion free $\mathbf{Z}_{(p)}$-algebra, F is canonically isomorphic to a formal group law F' which is *p-typical*, meaning that its logarithm has the form
$$\log_{F'}(x) = x + \sum_{n>0} \ell_n x^{p^n} \in R \otimes \mathbf{Q}[[x]].$$
The analog of the Lazard ring for such groups is much smaller than L, having polynomial generators v_n only in each dimension of the form $2(1-p^n)$.

The topological analog of Cartier's theorem is the splitting of $MU_{(p)}$ into a wedge of suspensions of spectra known as BP, originally constructed by Brown and Peterson [BP66a].

It is more convenient for us to work with the Adams-Novikov spectral sequence, which maps to the Adams spectral sequence. It has a family of elements in filtration 2, namely
$$\beta_{i/j} \in \mathrm{Ext}^{2,6i-2j}_{MU_*(MU)}(MU_*, MU_*)$$
for certain values of of i and j. When $j=1$, it is customary to omit it from the notation. The definition of these elements can be found in [Rav86, Chapter 5].

Here are the first few of these in the relevant bidegrees.
$$\begin{array}{llll} \theta_3: & \beta_{4/4} \text{ and } \beta_3 & \in & \mathrm{Ext}^{2,16}_{MU_*(MU)}(MU_*, MU_*), \\ \theta_4: & \beta_{8/8} \text{ and } \beta_{6/2} & \in & \mathrm{Ext}^{2,32}_{MU_*(MU)}(MU_*, MU_*), \\ \theta_5: & \beta_{16/16}, \beta_{12/4} \text{ and } \beta_{11} & \in & \mathrm{Ext}^{2,64}_{MU_*(MU)}(MU_*, MU_*), \end{array}$$
and so on. In the bidegree of θ_j, only $\beta_{2^{j-1}/2^{j-1}}$ has a nontrivial image (namely h_j^2) in the Adams spectral sequence. There is an additional element in this bidegree, namely $\alpha_1\alpha_{2^j-1}$. Thus
$$\mathrm{Ext}^{2,2^{j+1}}_{MU_*(MU)}(MU_*, MU_*)$$

is an elementary abelian 2-group of rank $[(j+3)/2]$. An element in it maps to h_j^2 in the corresponding classical Adams Ext group iff the coefficient of $\beta_{2^{j-1}/2^{j-1}}$ (with respect to the evident basis) is nontrivial.

We need to show that any such element has nontrivial image the Adams-Novikov spectral sequence for Ω.

Detection Theorem. *Let $j \geq 6$ and let $x \in \mathrm{Ext}_{MU_*(MU)}^{2,2^{j+1}}(MU_*, MU_*)$ be any element whose image in $\mathrm{Ext}_{\mathscr{A}}^{2,2^{j+1}}(\mathbf{Z}/2, \mathbf{Z}/2)$ is h_j^2. Then the image of x in $H^{2,2^{j+1}}(C_8; \pi_*\Omega_{\mathbb{O}})$ is nonzero.*

We will prove this by showing the same is true after we map the latter to a simpler Ext group defined in terms of another algebraic tool, *the theory of formal A-modules*, where A is the ring of integers in a suitable field. In that setting the group in the relevant bidegree is $(\mathbf{Z}/8)^4$. We will show that $\beta_{2^{j-1}/2^{j-1}}$ maps to an element of order 2, while the other basis elements map to 0.

7.2. Formal A-modules

Recall the a formal group law (see Sidebar 4) over a ring R is a power series

$$F(x,y) = x + y + \sum_{i,j>0} a_{i,j}x^iy^j \in R[[x,y]]$$

with certain properties.

For positive integers m one has power series $[m](x) \in R[[x]]$ defined recursively by $[1](x) = x$ and

$$[m](x) = F(x, [m-1](x)).$$

These satisfy

$$[m+n](x) = F([m](x), [n](x)) \text{ and } [m]([n](x)) = [mn](x).$$

With these properties we can define $[m](x)$ uniquely for all integers m, and we get a homomorphism

$$\begin{aligned} \tau : \mathbf{Z} &\to \mathrm{End}(F) \\ m &\mapsto [m](x) \end{aligned}$$

from the integers to the endomorphism ring of F.

If the ground ring R is an algebra over the p-local integers $\mathbf{Z}_{(p)}$ or the p-adic integers $\mathbf{Z}_p$, then we can make sense of $[m](x)$ for m in $\mathbf{Z}_{(p)}$ or $\mathbf{Z}_p$.

Now suppose R is an algebra over a larger ring A, such as the ring of localized integers in a number field or a finite extension of the p-adic numbers. We say that the formal group law F is a *formal A-module* if the homomorphism τ extends to A in such a way that

$$[a](x) \equiv ax \bmod (x^2) \text{ for } a \in A.$$

The theory of formal A-modules is well developed. Lubin-Tate [LT65] used it to do local class field theory.

The example of interest to us is $A = \mathbf{Z}_2[\zeta_8]$, where ζ_8 is a primitive 8th root of unity. It is the ring of integers in a cyclotomic extension of the 2-adic numbers $\mathbf{Q}_2$. The maximal ideal of A is generated by $\pi = \zeta_8 - 1$, and π^4 is a unit multiple of 2. There is a formal A-module F over $R_* = A[w^{\pm 1}]$ (with $|w| = 2$) satisfying

$$\log_F(F(x,y)) = \log_F(x) + \log_F(y)$$

where

$$\log_F(x) = \sum_{n \geq 0} \frac{w^{2^n-1}x^{2^n}}{\pi^n}. \tag{8}$$

We need to consider the cohomology of C_8 with coefficients in R_*, where the action of the group is trivial on A and $\gamma w = \zeta w$ for a group generator γ. $H^*(C_8; R_*)$ is the cohomology of the cochain complex

$$R_* \xrightarrow{\gamma-1} R_* \xrightarrow{\text{Trace}} R_* \xrightarrow{\gamma-1} \cdots$$

where Trace is multiplication by $1 + \gamma + \cdots + \gamma^7$.

The cohomology groups $H^s(C_8; R_*)$ for $s > 0$ are periodic in s with period 2. The part of it that matters to us is

$$H^2(C_8; R_{16m}) = w^{8m}A/(8) = w^{8m}\mathbf{Z}/(8)\left\{1,\, \pi,\, \pi^2,\, \pi^3\right\}.$$

7.3. The relation between $MU^{(C_8)}$ and formal A-modules

What does this have to do with our spectrum $\Omega_{\mathbb{O}} = D^{-1}MU^{(C_8)}$? Recall that D is given by $\bar{\mathfrak{d}}_1^{C_8} N_4^8(\bar{\mathfrak{d}}_2^{C_4}) N_2^8(\bar{\mathfrak{d}}_4^{C_2})$. We saw earlier that inverting a product of this sort is needed to get the Periodicity Theorem, but we did not explain the choice of subscripts of $\bar{\mathfrak{d}}$. They are the smallest ones that satisfy the second part of the following.

Lemma 7.1. *The classifying homomorphism $\lambda : \pi_*MU \to R_*$ for F factors through $\pi_*MU^{(C_8)}$ in such a way that*

- *the homomorphism $\lambda^{(4)} : \pi_*^u MU^{(C_8)} \to R_*$ is equivariant, where C_8 acts on $\pi_*MU^{(C_8)}$ as before, it acts trivially on A and $\gamma w = \zeta_8 w$ for a generator γ of C_8.*
- *The element $D \in \pi_*MU^{(C_8)}$ that we invert to get $\Omega_{\mathbb{O}}$ goes to a unit in R_*.*

We will sketch the proof of this later.

7.4. Deriving the Detection Theorem from Lemma 7.1

Lemma 7.1 implies that there is a diagram

$$
\begin{array}{ccccc}
 & S^0 & \longrightarrow & MU^{(C_8)} & \\
 & & & & \\
 & \mathrm{Ext}^{s,t}_{MU_*(MU)}(MU_*, MU_*) & \longrightarrow & \mathrm{Ext}^{s,t}_{MU_*(MU)}\left(MU_*, MU_*(MU^{(C_8)})\right) & \\
 & & & \| & \\
 & & & H^s(C_8; \pi^u_t MU^{(C_8)}) & \xrightarrow{\lambda^{(4)*}} \\
 & & & \downarrow & \\
H^s(C_8; \pi_t \Omega_{\mathbb{O}}) & = & & H^s(C_8; \pi_t D^{-1} MU^{(C_8)}) & \longrightarrow H^s(C_8; R_t)
\end{array}
$$

where the unit map at the top induces the homomorphism below it. We prove the Detection Theorem by showing that the composite map

$$
\begin{array}{ccc}
\mathrm{Ext}^{2,2^{j+1}}_{MU_*(MU)}(MU_*, MU_*) & \longrightarrow & H^2(C_8; R_{2^{j+1}}) \\
\| & & \| \\
\mathbf{Z}/2\left\{\alpha_1\alpha_{2^j-1},\ \beta_{2^{j-1}/2^{j-1}},\ \beta_{3\cdot 2^{j-3}/2^{j-3}},\ \dots\right\} & & w^{2^j} A/(8)
\end{array}
$$

sends $\beta_{2^{j-1}/2^{j-1}}$ to a nontrivial element of order 2, and each of the other basis elements to 0. This is a calculation requiring tools from BP-theory which is spelled out in [HHR, §11.4].

7.5. The proof of Lemma 7.1

To prove the first part, consider the following diagram for an arbitrary ring K.

$$
\begin{array}{ccccc}
 & & \pi^u_* MU^{(C_4)} & & \\
 & & \| & & \\
 & & \pi_* MU^{(2)} & & \\
 & & \| & & \\
 & & MU_*(MU) & & \\
 & \nearrow^{\eta_L} & \vdots\ \lambda^{(2)} & {}^{\eta_R}\nwarrow & \\
\pi_* MU & & \vdots & & \pi_* MU \\
 & \searrow_{\lambda_1} & \downarrow & {}_{\lambda_2}\swarrow & \\
 & & K & &
\end{array}
$$

The maps λ_1 and λ_2 classify two formal group laws F_1 and F_2 over K. The Hopf algebroid $MU_*(MU)$ represents strict isomorphisms between formal group laws. Hence the existence of $\lambda^{(2)}$ is equivalent to that of a strict isomorphism between F_1 and F_2.

Similarly consider the following diagram.

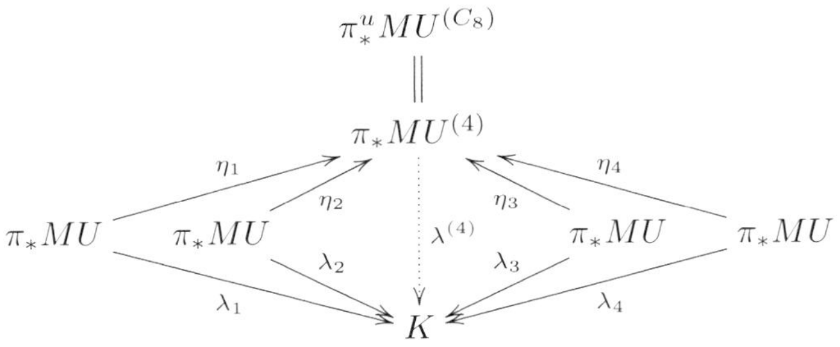

The four maps η_j are induced by the four unit maps from MU to $MU^{(4)}$, the spectrum underlying $MU^{(C_8)}$. The existence of $\lambda^{(4)}$ is equivalent to that of strict isomorphisms between the formal group laws F_j classified by the λ_j.

Now suppose that $K = R_*$ and each F_j is the formal group law given by (8). Then we have the required isomorphisms, so $\lambda^{(4)}$ exists. The inclusions η_j are related to the C_8-action on $\pi_* MU^{(4)}$ by

$$\gamma\eta_j = \eta_{j+1} \qquad \text{for } 1 \leq j \leq 3,$$

and $\gamma\eta_4$ differs from η_1 by $(-1)^i$ in dimension $2i$. The λ_j can be chosen so that

$$\gamma\lambda_j = \lambda_{j+1} \qquad \text{for } 1 \leq j \leq 3.$$

It follows that $\lambda^{(4)}$ is equivariant with respect to the C_8-actions on its source and target. This proves the first part of Lemma 7.1.

For the second part, recall that

$$D = N_2^8(\bar{r}_{15}^{C_2}\bar{r}_3^{C_4}\bar{r}_1^{C_8}).$$

The norm sends products to products, and $N(x)$ is a product of conjugates of x under the action of C_8. Hence its image in R_* is a unit multiple of that of a power of x, so it suffices to show that each of the three elements $\bar{r}_{15}^{C_2}$, $\bar{r}_3^{C_4}$ and $\bar{r}_1^{C_8}$ maps to a unit in R_*. We refer the reader to [HHR, §11.5] for the details.

8. The Slice and Reduction theorems

Recall that a pivotal step in our proof is the Slice Theorem, which identifies the layers in the slice tower for $MU_{\mathbf{R}}$ and its relatives. In [HHR, §6] it is derived from the Reduction Theorem, which we now state.

For each cyclic 2-group $G = C_{2^{n+1}}$ there is a certain equivariant (noncommutative) ring spectrum A, defined in (3), which is a certain wedge of slice cells. It maps to $MU^{(G)}$ in such a way that the underlying wedge of spheres hits all of the underlying homotopy of $MU^{(G)}$. We can map it to S^0 (with trivial group action) by sending all of the 0-connected summands to a point. Thus both $MU^{(G)}$ and S^0 are A-modules.

Reduction Theorem. *Let* $R(\infty)^{(G)} = MU^{(G)} \wedge_A S^0$. *The map*

$$f_G : R(\infty)^{(G)} \to H\mathbf{Z}_{(2)}$$

given by the 0*th slice is an equivariant equivalence.*

Recall that we have localized MU at the prime 2. We will denote $R(\infty)^{(G)}$ and f_G simply by $R(\infty)$ and f when the value of G is clear from the context.

The nonequivariant version of this is obvious. It says that if we kill of all of the positive dimensional homotopy of the spectrum underlying $MU^{(G)}$, we get $H\mathbf{Z}_{(2)}$, its zeroth Postnikov section. The case $G = C_2$ was proved by Hu-Kriz in [HK01].

The derivation of the Slice Theorem from the Reduction Theorem is essentially formal.

The proof of the Reduction Theorem is the hardest calculation in our paper and is treated in [HHR, §7].

8.1. Deriving the Slice Theorem from the Reduction Theorem

It is possible to define an A-module structure on M_d, the wedge of all $(d-1)$-connected summands of A. Then $M_{2d-1} = M_{2d}$ and $\widehat{W}_{2d}$ (defined in 3.5) is the cofiber of the inclusion $M_{2d+2} \to M_{2d}$, which is also an A-module.

Let

$$K_d = MU^{(G)} \wedge_A M_d \qquad \text{and} \qquad \tilde{P}^d = MU^{(G)}/K_{d-1}$$

Then it is not difficult to show that the fiber of the inclusion $\tilde{P}^{2d} \to \tilde{P}^{2d-1}$ is $R(\infty) \wedge \widehat{W}_{2d}$. The Reduction Theorem implies that this is $H\mathbf{Z}_{(2)} \wedge \widehat{W}_{2d}$. It also enables us to identify the diagram

$$\begin{array}{ccccccc}
\cdots \longrightarrow & \tilde{P}^{2d+1} & \longrightarrow & \tilde{P}^{2d} & \longrightarrow & \tilde{P}^{2d-1} & \longrightarrow \cdots \\
& \uparrow & & \uparrow & & \uparrow & \\
& * & & R(\infty) \wedge \widehat{W}_{2d} & & * & \\
& & & \| & & & \\
& & & \wedge \widehat{W}_{2d} & & &
\end{array}$$

with the slice tower (2) for $MU^{(G)}$, i.e., to show that $\tilde{P}^m$ is equivalent to $P^m MU^{(G)}$. The Slice Theorem follows.

There is a converse [HHR, Prop. 6.9] which says roughly that if a ring spectrum R (such as $MU^{(G)} \wedge MU^{(G)}$) with properties similar to those of $MU^{(G)}$ satisfies the Slice Theorem, then the Reduction Theorem follows. Since this converse is used in the proof of the Reduction Theorem itself, we will state it more precisely.

For a finite cyclic 2-group G, let R be a connective 2-local G-ring spectrum with $\pi_*^u R$ a polynomial algebra over $\mathbf{Z}_{(2)}$ on positive dimensional generators. We suppose further that $\pi_*^u R$ is refined by a map from a spectrum A of the form

$$A = N_2^g S^0[\overline{x}_1, \overline{x}_2, \dots]$$

where $\overline{x}_i \in \pi^{C_2}_{k_i\rho_2}X$ with $k_i > 0$ for each i. Then we can define M_d, K_d and $\tilde{P}^d$ for R as we did for $MU^{(G)}$ above. In particular $\tilde{P}^0$ would be the analog of $R(\infty)$ as above.

Then similar arguments to those above give

Proposition 8.1. *For R as above, if $\tilde{P}^0$ is equivalent to $H\mathbf{Z}_{(2)}$, then R is pure and isotropic.*

Here is our converse.

Proposition 8.2. *Let R be a connective 2-local G-ring spectrum as above, and suppose that it is pure (Definition 3.2). Then the tower $\left\{\tilde{P}^d\right\}$ is the slice tower for R. In particular, $\tilde{P}^0$ is $P^0R = H\mathbf{Z}_{(2)}$.*

8.2. The proof of the Reduction Theorem

We will use a double induction argument. Assuming that the map f_G (which is defined only for nontrivial G) is H-equivariant for a subgroup $H \subseteq G$, we will show that

(i) When H is a proper subgroup of G, it can be replaced by the next larger subgroup of G.

(ii) G can be replaced by the next larger supergroup G' of G.

We have the statement for trivial H and any G since that is the nonequivariant case. Hence the Reduction Theorem will follow from the two inductive steps.

The second inductive step follows immediately from the fact that the restriction of $R(\infty)^{(G')}$ to G is $R(\infty)^{(G)}$. This means that we need only show the first step in the case where H is the index two subgroup of G.

For A as in (3) and an integer $i \geq 0$ we define an A-module

$$R(i)^{(G)} = MU^{(G)} \wedge_A A'_i \qquad \text{where} \qquad A'_i = N_2^g S^0[\overline{r}_{i+1}, \overline{r}_{i+2}, \ldots]. \tag{9}$$

Roughly speaking, $R(i)^{(G)}$ is obtained from $MU^{(G)}$ by killing the subring of the underlying homotopy generated by elements below dimension $2i$. We will often denote it simply by $R(i)$.

Here are three consequences of the inductive hypothesis, proved in [HHR, Propositions 7.2 and 7.3].

Proposition 8.3. *Let $H \subset G$ be the index 2 subgroup. If the inductive hypothesis above holds in this case, then*

(i) The Slice Theorem holds for H when $|H| > 1$.

(ii) $i_H^* MU^{(G)}$ is a pure isotropic spectrum.

(iii) $i_H^* R(i)^{(G)}$ (see (9)) is pure and isotropic.

Proof. (i) Since $i_H^* R(\infty)^{(G)} = R(\infty)^{(H)}$, the hypothesis of this proposition is the same as the Reduction Theorem for H. It implies the Slice Theorem for H.

(ii) $MU^{(G)}$ and $MU^{(H)} \wedge MU^{(H)}$ are underlain by the same smash power of MU. Thus $i_H^* MU^{(G)}$ is a wedge of suspensions of $MU^{(H)}$, and this splitting can be shown to be equivariant. Hence $i_H^* MU^{(G)}$ is pure isotropic because $MU^{(H)}$ is.

(iii) $i_H^* R(i)^{(G)}$ has an equivariant splitting similar to that of $i_H^* MU^{(G)}$ in which each summand is a suspension of $R(i)^{(H)}$. The reduction theorem for H implies that $R(i)^{(H)}$ is pure and isotropic by Proposition 8.1, so the same is true of $i_H^* R(i)^{(G)}$. □

Now we smash the map of the Reduction Theorem with the isotropy separation sequence of (4) and get a diagram

$$\begin{array}{ccccc} E\mathcal{P}_+ \wedge R(\infty) & \longrightarrow & R(\infty) & \longrightarrow & \tilde{E}\mathcal{P} \wedge R(\infty) \\ \downarrow{\scriptstyle f'} & & \downarrow{\scriptstyle f} & & \downarrow{\scriptstyle f''} \\ E\mathcal{P}_+ \wedge H\mathbf{Z}_{(2)} & \longrightarrow & H\mathbf{Z}_{(2)} & \longrightarrow & \tilde{E}\mathcal{P} \wedge H\mathbf{Z}_{(2)} \end{array}$$

We want to show that f is an equivariant equivalence. The inductive hypothesis implies that f' is one, so it suffices to show that f'' is one. We see from Theorem 5.2 (iii) this boils down to showing that the map

$$\pi_* \Phi^G f = \pi_*^G f'' : \pi_* \Phi^G R(\infty) \to \pi_* \Phi^G H\mathbf{Z}_{(2)} \tag{10}$$

is an isomorphism.

It is not hard to show that the source and target of (10) are abstractly isomorphic and of finite type [HHR, Prop. 7.5].

Proposition 8.4.

$$\pi_* \Phi^G H\mathbf{Z}_{(2)} = \mathbf{Z}/2[b] \quad \textit{where} \quad b = u_{2\sigma} a_\sigma^{-2} \in \pi_2 \Phi^G H\mathbf{Z}_{(2)} \subset a_\sigma^{-1} \pi_\star^G H\mathbf{Z}_{(2)}.$$

and

$$\pi_n \Phi^G R(\infty) = \begin{cases} \mathbf{Z}/2 & \textit{for } n \geq 0 \textit{ and even} \\ 0 & \textit{otherwise.} \end{cases}$$

Note that $\pi_* \Phi^G R(\infty)$ lacks a multiplication because it is not obvious that $R(\infty)^{(G)}$ is a ring spectrum. We only know it is as a consequence of the Reduction Theorem.

Proof. The first assertion follows from computations similar to those of 4.2.

For the second we will compute $\pi_* \Phi^G R(i)$ by induction on i. Since $R(0) = MU^{(G)}$, $\Phi^G R(0) = MO$, so

$$\pi_* \Phi^G R(0) = \pi_* MO = \mathbf{Z}/2[h_j, j \neq 2^k - 1]$$

where $h_j \in \pi_j MO$. We can identify the fiber of the map $\Phi^G R(i-1) \to \Phi^G R(i)$ and we have a cofiber sequence

$$\Sigma^k \Phi^G R(i-1) \xrightarrow{\Phi^G \overline{r}_i} \Phi^G R(i-1) \longrightarrow \Phi^G R(i)$$

where

$$\Phi^G \overline{r}_i = \begin{cases} h_i & \text{for } i \neq 2^k - 1 \\ 0 & \text{for } i = 2^k - 1. \end{cases} \tag{11}$$

The indicated value of $\pi_* \Phi^G R(\infty)$ follows. □

Thus it suffices to show that the homomorphism of (10) is onto. This job is simplified by the following, which is [HHR, Lemma 7.7].

Let

$$B_i = N_2^g S^0[\overline{r}_i] \quad \text{and} \quad X_i = MU^{(G)} \wedge_{B_i} S^0. \tag{12}$$

X_i is the spectrum obtained from $MU^{(G)}$ by killing the polynomial generators in dimension $2i$ of its underlying homotopy.

Lemma 8.5. *If for every $k \geq 1$, the class $b^{2^{k-1}}$ is in the image of*

$$\pi_{2^k} \Phi^G MU^{(G)} \wedge_{B_{2^{k-1}}} S^0 \to \pi_{2^k} \Phi^G H\mathbf{Z}_{(2)},$$

then the map of (10) is onto.

Proof. We have

$$R(\infty) = X_1 \wedge_{MU^{(G)}} X_2 \wedge_{MU^{(G)}} \cdots$$

This implies that if $b^{2^{k-1}}$ is in the image for each $k \geq 1$, then every power of b is. □

Let $c_k = 2^k - 1$. To show that the hypothesis of Lemma 8.5 is met we need to study the element

$$N_2^g \overline{r}_{c_k} \in \pi^G_{c_k \rho_G} MU^{(G)},$$

which we will see to be related to an element mapping to $b^{2^{k-1}}$ under the homomorphism of Lemma 8.5. Its image in $\pi^G_{c_k\rho_G} \tilde{E}\mathcal{P} \wedge X_{c_k}$ is zero for two reasons: we are coning it off to form X_{c_k}, and it is killed by the functor Φ^G. We will verify the hypothesis of Lemma 8.5 by chasing this element around the diagram

$$\begin{array}{ccccc} E\mathcal{P}_+ \wedge MU^{(G)} & \longrightarrow & MU^{(G)} & \longrightarrow & \tilde{E}\mathcal{P} \wedge MU^{(G)} \\ \downarrow & & \downarrow & & \downarrow \\ E\mathcal{P}_+ \wedge X_{c_k} & \longrightarrow & X_{c_k} & \longrightarrow & \tilde{E}\mathcal{P} \wedge X_{c_k} \\ \downarrow & & \downarrow & & \downarrow \\ E\mathcal{P}_+ \wedge H\mathbf{Z}_{(2)} & \longrightarrow & H\mathbf{Z}_{(2)} & \longrightarrow & \tilde{E}\mathcal{P} \wedge H\mathbf{Z}_{(2)} \end{array} \tag{13}$$

Here the rows are cofiber sequences, but the columns are not.

Since $S^{i\rho_G}$ is obtained from S^i by attaching induced cells, one can show [HHR, Remark 2.34] that the restriction map

$$\pi^G_{i\rho_G} \tilde{E}\mathcal{P} \wedge X \to \pi^G_i \tilde{E}\mathcal{P} \wedge X$$

is an isomorphism for any G-spectrum X. The image of $N_2^g \overline{r}_{c_k}$ in

$$\pi^G_{c_k\rho_G} \tilde{E}\mathcal{P} \wedge MU^{(G)} \approx \pi^G_{c_k} \tilde{E}\mathcal{P} \wedge MU^{(G)} \approx \pi_{c_k} MO$$

is zero by (11). Thus in the top row of (13), $N_2^g \overline{r}_{c_k}$ has a preimage

$$y_k \in \pi^G_{c_k\rho_G} E\mathcal{P}_+ \wedge MU^{(G)}.$$

Proposition 8.6. *Any choice of y_k as above has nontrivial image in*

$$\pi^G_{c_k\rho_G} E\mathcal{P}_+ \wedge H\mathbf{Z}_{(2)}.$$

Proof of the Reduction Theorem modulo Proposition 8.6. We have already seen that the Reduction Theorem follows from the hypothesis of Lemma 8.5, that b^{c_k} is in the image of the map induced by f. The image of y_k in $\pi^G_{c_k\rho_G} E\mathcal{P}_+ \wedge X_{c_k}$ has trivial image in $\pi^G_{c_k\rho_G} X_{c_k}$, so it has a preimage

$$\tilde{y}_k \in \pi^G_{c_k\rho_G+1} \tilde{E}\mathcal{P}_+ \wedge X_{c_k}.$$

The image of $\tilde{y}_k$ in $\pi^G_{c_k\rho_G+1} \tilde{E}\mathcal{P}_+ \wedge H\mathbf{Z}_{(2)}$ is nontrivial by Proposition 8.6.

Now consider the diagram

$$\begin{array}{ccc} \pi^G_{c_k\rho_G+1} \tilde{E}\mathcal{P}_+ \wedge X_{c_k} & \xrightarrow{\approx} & \pi^G_{2^k} \tilde{E}\mathcal{P}_+ \wedge X_{c_k} \\ \downarrow & & \downarrow \\ \pi^G_{c_k\rho_G+1} \tilde{E}\mathcal{P}_+ \wedge H\mathbf{Z}_{(2)} & \xrightarrow{\approx} & \pi^G_{2^k} \tilde{E}\mathcal{P}_+ \wedge H\mathbf{Z}_{(2)} \end{array}$$

where the horizontal maps are restrictions along the inclusion $S^{2^k} \to S^{c_k\rho_G+1}$. The lower right group is $\mathbf{Z}/2$ generated by $b^{2^{k-1}}$. We have just shown that $\tilde{y}_k$ maps nontrivially to the lower left group, so $b^{2^{k-1}}$ is in the image of the map induced by f. □

Proposition 8.6 concerns the map

$$E\mathcal{P}_+ \wedge MU^{(G)} \to E\mathcal{P}_+ \wedge H\mathbf{Z}_{(2)}.$$

Each cell of both source and target is induced up from the index 2 subgroup H, so we can use the inductive hypothesis. We will use the factorization

$$E\mathcal{P}_+ \wedge MU^{(G)} \to E\mathcal{P}_+ \wedge R(2^k-2)^{(G)} \to E\mathcal{P}_+ \wedge H\mathbf{Z}_{(2)},$$

with $R(2^k-2)^{(G)}$ as in (9), and replace y_k by its image in $\pi^G_{c_k\rho_G} E\mathcal{P}_+ \wedge R(2^k-2)^{(G)}$. We will study the smash product of $E\mathcal{P}_+$ with the slice tower for $R(2^k-2)^{(G)}$. The following is [HHR, Lemma 7.14].

Lemma 8.7. *For $0 < m < c_k g$,*

$$\pi^G_{c_k\rho_G} E\mathcal{P}_+ \wedge P^m_m R(2^k-2)^{(G)} = 0$$

and there is an exact sequence

$$\begin{array}{ccc} \pi^G_{c_k\rho_G} E\mathcal{P}_+ \wedge P_{c_k g} R(2^k-2)^{(G)} & \longrightarrow & \pi^G_{c_k\rho_G} E\mathcal{P}_+ \wedge R(2^k-2)^{(G)} \\ & & \downarrow \\ & & \pi^G_{c_k\rho_G} E\mathcal{P}_+ \wedge H\mathbf{Z}_{(2)} = \mathbf{Z}/2 \end{array}$$

Proof. The smash product of $E\mathcal{P}_+$ with the slice tower for $R(2^k-2)^{(G)}$ is determined by the restriction of the latter to H, which has the expected properties by induction. The result follows from an examination of its equivariant homotopy groups. □

The exact sequence of Lemma 8.7 shows that y_k has nontrivial image in the group $\pi^G_{c_k\rho_G} E\mathcal{P}_+ \wedge H\mathbf{Z}_{(2)}$ iff it is not in the image of $\pi^G_{c_k\rho_G} E\mathcal{P}_+ \wedge P_{c_k g} R(2^k-2)^{(G)}$. For this we will identify a property of this image not shared by y_k. The following is [HHR, Prop. 7.15].

Proposition 8.8. *The image of*

$$\pi^G_{c_k\rho_G} E\mathcal{P}_+ \wedge P_{c_k g} R(2^k-2)^{(G)} \longrightarrow \pi^G_{c_k\rho_G} R(2^k-2)^{(G)} \xrightarrow{i_0^*} \pi^u_{c_k g} R(2^k-2)^{(G)}$$

is contained in the image of $(\gamma-1)$*, where* γ *is a generator of* G *and* i_0 *denotes restriction to the trivial group.*

The class y_k does not have this property because its image in $\pi^u_{c_k g} R(2^k-2)^{(G)}$ is $i_0^* N_2^g \bar{r}_{c_k}$, which generates a summand isomorphic to the sign representation of G. Hence this result implies the Reduction Theorem. For its proof we refer the reader to the last two pages of [HHR, §7].

References

[Ada58] J. F. Adams, On the structure and applications of the Steenrod algebra, *Comment. Math. Helv.*, **32** (1958), 180–214.

[Ada60] J. F. Adams, On the non-existence of elements of Hopf invariant one, *Ann. of Math. (2)*, **72** (1960), 20–104, 1960.

[Ada62] J. F. Adams, Vector fields on spheres, *Ann. of Math. (2)*, **75** (1962), 603–632.

[Ada66] J. F. Adams, On the groups $J(X)$, IV, *Topology*, **5** (1966), 21–71.

[Arf41] Cahit Arf, Untersuchungen über quadratische Formen in Körpern der Charakteristik 2, I, *J. Reine Angew. Math.*, **183** (1941), 148–167.

[Ati66] M. F. Atiyah, K-theory and reality, *Quart. J. Math. Oxford Ser. (2)*, **17** (1966), 367–386.

[BJM84] M. G. Barratt, J. D. S. Jones and M. E. Mahowald, Relations amongst Toda brackets and the Kervaire invariant in dimension 62, *J. London Math. Soc. (2)*, **30** (1984), no. 3, 533–550.

[BMT70] M. G. Barratt, M. E. Mahowald and M. C. Tangora, Some differentials in the Adams spectral sequence, II, *Topology*, **9** (1970), 309–316.

[Bot59] R. Bott, The stable homotopy of the classical groups, *Annals of Mathematics*, **70** (1969) 313–337.

[BP66a] E. H. Brown and F. P. Peterson, A spectrum whose $\mathbf{Z}_p$ cohomology is the algebra of reduced p-th powers, *Topology*, **5** (1966), 149–154.

[BP66b] E. H. Brown, Jr. and F. P. Peterson, The Kervaire invariant of $(8k+2)$-manifolds, *Amer. J. Math.*, **88** (1966), 815–826.

[Bro69] W. Browder, The Kervaire invariant of framed manifolds and its generalization, *Ann. of Math. (2)*, **90** (1969), 157–186.

[Car67] P Cartier, Modules associés à un groupe formel commutatif, Courbes typiques, *C. R. Acad. Sci. Paris Sér. A-B*, **265** (1967), A129–A132.

[Far96] E. Dror Farjoun, *Cellular spaces, null spaces and homotopy localization*, Lecture Notes in Mathematics, 1622, Springer-Verlag, New York, 1996.

[Fre38] H. Freudenthal, Über die Klassen der Sphärenabbildungen I, Grosse Dimnesionionen , *Compositio Math.*, **5** (1938), 299–314.

[GM95] J. P. C. Greenlees and J. P. May, Equivariant stable homotopy theory, in *Handbook of algebraic topology*, 277–323, North-Holland, Amsterdam, 1995.

[Haz78] M. Hazewinkel, *Formal groups and applications*, Pure and Applied Mathematics, 78, Academic Press, New York, 1978.

[HHR] M. A. Hill, M. J. Hopkins and D. C. Ravenel, The non-existence of elements of Kervaire invariant one, arXiv preprint, arXiv:0908.3724v2.

[Hop35] H. Hopf, Über die Abbildungen von Sphar̃en auf Sphären niedrigerer Dimension, *Fundamenta Mathematica*, **25** (1935), 427–440.

[HK01] P. Hu and I. Kriz, Real-oriented homotopy theory and an analogue of the Adams-Novikov spectral sequence, *Topology*, **40** (2001), no. 2, 317–399.

[Jam55] I. M. James, Reduced product spaces, *Ann. of Math. (2)*, **62** (1955), 170–197.

[Jam56a] I. M. James, On the suspension triad, *Ann. of Math. (2)*, **63** (1956), 191–247.

[Jam56b] I. M. James, The suspension triad of a sphere, *Ann. of Math. (2)*, **63** (1956), 407–429.

[Jam57] I. M. James, On the suspension sequence, *Ann. of Math. (2)*, **65** (1957), 74–107.

[Jon78] J. D. S. Jones, The Kervaire invariant of extended power manifolds, *Topology*, **17** (1978), no. 3, 249–266.

[Ker60] M. A. Kervaire, A manifold which does not admit any differentiable structure, *Comment. Math. Helv.*, **34** (1960), 257–270.

[Laz55] M. Lazard, Lois de groupes et analyseurs, *Ann. Écoles Norm. Sup.*, **72** (1955), 299–400.

[LT65] J. Lubin and J. Tate, Formal complex multiplication in local fields, *Ann. of Math. (2)*, **81** (1965), 380–387.

[Mah67] M. Mahowald, *The metastable homotopy of* S^n, Memoirs of the American Mathematical Society, No. 72, American Mathematical Society, Providence, RI, 1967.

[Mah82] M. Mahowald, The image of J in the EHP sequence, *Ann. of Math. (2)*, **116** (1982), no. 1, 65–112.

[Mil60] J. W. Milnor, On the cobordism ring Ω^* and a complex analogue. I, *Amer. J. Math.*, **82** (1960), 505–521.

[MS74] J. W. Milnor and J. D. Stasheff, *Characteristic classes*, Annals of Mathematics Studies, No. 76, Princeton University Press, Princeton, NJ, 1974.

[MT67] M. Mahowald and M. Tangora, Some differentials in the Adams spectral sequence, *Topology*, **6** (1967), 349–369.

[Nov60] S. P. Novikov, Some problems in the topology of manifolds connected with the theory of Thom spaces, *Soviet Math. Dokl.*, **1** (1960), 717–720.

[Nov62] S. P. Novikov, Homotopy properties of Thom complexes, *Mat. Sb. (N.S.)*, **57 (99)** (1962), 407–442.

[Nov67] S. P. Novikov, Methods of algebraic topology from the point of view of cobordism theory, *Izv. Akad. Nauk SSSR Ser. Mat.*, **31** (1967), 855–951.

[Pon38] L. S. Pontryagin, A classification of continuous transformations of a complex into a sphere, 1 and 2, *Doklady Akad. Nauk SSSR (N.S.)*, **19** (1938), 147–149 and 361–363.

[Pon50] L. S. Pontryagin, Homotopy classification of the mappings of an $(n+2)$-dimensional sphere on an n-dimensional one, *Doklady Akad. Nauk SSSR (N.S.)*, **70** (1950), 957–959.

[Pon55] L. S. Pontryagin, *Gladkie mnogoobraziya i ih primeneniya v teorii gomotopiĭ[Smooth manifolds and their applications in homotopy theory]*. Trudy Mat. Inst. im. Steklov. no. 45. Izdat. Akad. Nauk SSSR, Moscow, 1955.

[Qui69] D. Quillen, On the formal group laws of unoriented and complex cobordism theory, *Bull. Amer. Math. Soc.*, **75** (1969), 1293–1298.

[Rav86] D. C. Ravenel, *Complex cobordism and stable homotopy groups of spheres*, Pure and Applied Mathematics, 121, Academic Press Inc., Orlando, FL, 1986, errata and second edition available online at author's home page.

[Rav04] D. C. Ravenel, *Complex Cobordism and Stable Homotopy Groups of Spheres, Second Edition*, American Mathematical Society, Providence, 2004, available online at author's home page.

[Sna09] V. P. Snaith, *Stable homotopy around the Arf-Kervaire invariant*, Progress in Mathematics, 273, Birkhäuser Verlag, Basel, 2009.

[Sto68] R. E. Stong, *Notes on Cobordism Theory*, Mathematical Notes, Princeton University Press, Princeton, 1968.

[Tho54] R. Thom, Quelques propriétés globales des variétés différentiables, *Comment. Math. Helv.*, **28** (1954), 17–86.

[Whi42] G. W. Whitehead, On the homotopy groups of spheres and rotation groups, *Annals of Mathematics*, **43** (1942), 634–640.

[Whi50] G. W. Whitehead, The $(n+2)^{\mathrm{nd}}$ homotopy group of the n-sphere, *Ann. of Math. (2)*, **52** (1950), 245–247.

Department of Mathematics, University of Virginia, Charlottesville, VA 22904
E-mail address: michael.a.hill@math.uva.edu

Department of Mathematics, Harvard University, Cambridge, MA 02138
E-mail address: mjh@math.harvard.edu

Department of Mathematics, University of Rochester, Rochester, NY 14627
E-mail address: doug+pi@math.rochester.edu

Proceedings of 17th Gökova
Geometry-Topology Conference
pp. 64 – 78

Degree one cohomology with twisted coefficients of the mapping class group

Jørgen Ellegaard Andersen and Rasmus Villemoes

Abstract. Let Γ be the mapping class group of an oriented surface Σ of genus g with r boundary components. We prove that the first cohomology group of the mapping class group $H^1(\Gamma, \mathcal{O}(\mathcal{M}_{\mathrm{SL}_2(\mathbb{C})})^*)$ is non-trivial, where the coefficient module is the dual of the space of algebraic functions on the $\mathrm{SL}_2(\mathbb{C})$ moduli space over Σ.

1. Introduction

Let $\Gamma = \Gamma_{g,r}$ denote the mapping class group of a compact surface $\Sigma = \Sigma_{g,r}$ with genus g and r boundary components. There is an action of Γ on the moduli space $\mathcal{M}_G$ of flat G-connections over Σ. The vector space $\mathcal{O}(\mathcal{M}_G) \subseteq \mathrm{Fun}(\mathcal{M}_G, \mathbb{C})$ of regular functions on the moduli space is naturally a Γ-module.

Presently, we consider the special case of $G = \mathrm{SL}_2(\mathbb{C})$, and we simply write $\mathcal{M}$ for $\mathcal{M}_{\mathrm{SL}_2(\mathbb{C})}$. In this case, there is an isomorphism of Γ-modules

$$\nu\colon \mathcal{B}(\Sigma) \to \mathcal{O}(\mathcal{M}), \tag{1}$$

where the source denotes the complex vector space freely spanned by the set of multicurves on Σ, i.e., isotopy classes of a closed 1-submanifold. We usually think of this as a finite collection of pairwise non-intersecting, non-trivial, unoriented simple loops on Σ. Note that we explicitly allow a multicurve to contain isotopic connected components, and that it may also contain components parallel to a boundary component of Σ. Letting $B = B(\Sigma)$ denote the set of multicurves on Σ, $\mathcal{B} = \mathcal{B}(\Sigma)$ is simply the complex vector space spanned by B. There is a natural algebra structure on this space; for details on this see [1] and [2]. The isomorphism ν is given on a single simple loop γ by $\nu(\gamma) = -f_{\vec{\gamma}}$, where $\vec{\gamma}$ is any of the two oriented versions of γ, and $f_{\vec{\gamma}}$ is the function which to a gauge equivalence class $[A]$ of flat connections associates the trace of the holonomy of A along $\vec{\gamma}$.

We may think of $\mathcal{B}$ as the set of maps $B \to \mathbb{C}$ which vanish except for a finite number of multicurves. This is naturally embedded in the larger module of *all* maps $\hat{\mathcal{B}} = \mathrm{Map}(B, \mathbb{C})$; this is clearly the same as the algebraic dual $\mathcal{O}(\mathcal{M})^*$ of $\mathcal{O}(\mathcal{M})$. The action of Γ splits B into orbits. Let S denote a set of representatives of these orbits, and for $D \in S$,

2010 *Mathematics Subject Classification.* Primary 57M60; Secondary 20J06.
Key words and phrases. mapping class group, group cohomology, moduli spaces, multicurves.
Supported by QGM (Centre for Quantum Geometry of Moduli Spaces, funded by the Danish National Research Foundation).

let $\hat{M}_D$ (respectively M_D) denote the space of all maps from the orbit through D to $\mathbb{C}$ (respectively, the maps $\Gamma D \to \mathbb{C}$ which vanish for all but a finite number of multicurves in the orbit). With this notation, we obtain splittings of $\mathcal{B}$ and $\hat{\mathcal{B}}$ as Γ-modules

$$\mathcal{O}(\mathcal{M}) \cong \mathcal{B} \cong \bigoplus_{D\in S} M_D$$
$$\mathcal{O}(\mathcal{M})^* \cong \hat{\mathcal{B}} \cong \prod_{D\in S} \hat{M}_D \tag{2}$$

which induce decompositions in cohomology

$$H^*(\Gamma, \mathcal{B}) \cong \bigoplus_{D\in S} H^*(\Gamma, M_D) \tag{3}$$
$$H^*(\Gamma, \hat{\mathcal{B}}) \cong \prod_{D\in S} H^*(\Gamma, \hat{M}_D). \tag{4}$$

The isomorphism (3) depends on the well-known fact that Γ is finitely generated (see e.g. [3]). A cocycle $u\colon \Gamma \to \mathcal{O}(\mathcal{M})^* = \hat{\mathcal{B}} = \mathrm{Map}(B, \mathbb{C})$ may also be thought as a map $u\colon \Gamma \times B \to \mathbb{C}$ by simply putting $u(\gamma)(E) = u(\gamma, E)$.

Theorem 1.1. *A cocycle $u\colon \Gamma \to \mathcal{O}(\mathcal{M})^* = \hat{\mathcal{B}} = \mathrm{Map}(B, \mathbb{C})$ is a coboundary if and only if for each $D \in S$, the restriction of u to $\Gamma_D \times \{D\}$ is identically 0, where Γ_D denotes the stabilizer of the multicurve D in Γ.*

We will use this theorem to arrive at the main result:

Theorem 1.2. *For every $g, r \geq 0$, the cohomology group $H^1(\Gamma_{g,r}, \mathcal{O}(\mathcal{M})^*)$ is a direct product of summands $H^1(\Gamma, \hat{M}_D)$, each of which is finite-dimensional. Here D runs over a set of representatives of multicurves on Σ.*

In particular, we obtain by explicit examples

Corollary 1.3. *Whenever $\Gamma_{g,r}$ is non-trivial (i.e., when (g, r) is not $(0, 0)$ or $(0, 1)$), $H^1(\Gamma_{g,r}, \mathcal{O}(\mathcal{M})^*)$ is non-trivial.*

The motivation to study the cohomology of the mapping class group with these coefficients came from [4], particularly Proposition 6, where integrability of certain cocycles turns out to be an obstruction to finding a Γ-invariant equivalence between two equivalent star products on the moduli space. The motivation for studying that problem comes from the expectation that the star products discussed in [4] are equivalent to the star product which is constructed in [1] and which is the same as the ones induced on the $\mathrm{SL}_2(\mathbb{C})$-moduli space from the constructions given in [5] and [6]. For the first results in this direction please see [7].

The main result of this paper, Theorem 1.2, should be compared to the following vanishing results also due to the authors of this paper. The first of these concerns the same moduli space as considered in this paper, with coefficients in the space of algebraic functions.

Theorem 1.4. *For $g \geq 2$ and $r \geq 0$, the cohomology group $H^1(\Gamma, \mathcal{O}(\mathcal{M}_{\mathrm{SL}_2(\mathbb{C})}))$ vanishes.*

This result is proved in [8].

In another paper [9], we considered the abelian moduli space of a surface with at most one boundary component, i.e., the space

$$\mathcal{M}_{\mathrm{U}(1)} = \mathrm{Hom}(\pi_1\Sigma, \mathrm{U}(1)) \cong \mathrm{U}(1)^{2g}$$

of flat U(1)-connections on Σ. The mapping class group acts by measure-preserving diffeomorphisms of $\mathcal{M}_{\mathrm{U}(1)}$, so both $C^\infty(\mathcal{M}_{\mathrm{U}(1)})$ and $L^2(\mathcal{M}_{\mathrm{U}(1)})$ are Γ-modules. When the genus is at least 3, we obtained these results:

Theorem 1.5. *The cohomology group $H^1(\Gamma, L^2(\mathcal{M}_{\mathrm{U}(1)}))$ vanishes.*

Theorem 1.6. *The cohomology group $H^1(\Gamma, C^\infty(\mathcal{M}_{\mathrm{U}(1)}))$ vanishes.*

Also in [9] we derived a general result which applies to any unitary representation of the mapping class group.

Theorem 1.7. *Assume $g \geq 3$. Let $\Gamma \to \mathrm{U}(V)$ be a unitary representation of Γ on a real or complex Hilbert space V. For a simple closed curve γ on Σ, let $V_\gamma = V^{\tau_\gamma}$ denote the closed subspace fixed under the twist τ_γ and let $p_\gamma \colon V \to V_\gamma$ be the orthogonal projection. Then, for any cocycle $u \colon \Gamma \to V$, we have $p_\gamma u(\tau_\gamma) = 0$.*

This result should be understood in light of the observation that if u is the coboundary of some element $v \in V$, then $u(\tau_\gamma) = (1 - \tau_\gamma)v$ is killed by p_γ. Hence this necessary condition for the vanishing of $H^1(\Gamma, V)$ is always satisfied whenever $g \geq 3$.

This paper is organized as follows. In Section 2 we develop some of the basic properties of group cohomology which are needed in the calculations, ending with a proof of Theorem 1.1. In Section 3, we develop an algorithm to compute $H^1(\Gamma, \hat{M}_D)$ for any multicurve, which enables us to prove Theorem 1.2. This is used in Section 4 to give a generic example of a multicurve for which the cohomology is non-zero, hence proving Corollary 1.3.

2. Group cohomological background

If Γ is a group and A is a Γ-module, a *cocycle* is a map $u \colon \Gamma \to A$ satisfying the *cocycle condition*

$$u(gh) = u(g) + gu(h) \tag{5}$$

for all $g, h \in \Gamma$. A *coboundary* is a cocycle of the form $g \mapsto a - ga = \delta a(g)$ for some $a \in A$. The cohomology group $H^1(\Gamma, A)$ is the space of cocycles modulo the space of coboundaries.

Theorem 2.1 (Shapiro's Lemma). *Let H be a subgroup of Γ and A a left H-module. Then there are isomorphisms*

$$H_*(H, A) \cong H_*(\Gamma, \mathrm{Ind}_H^\Gamma A) \tag{6}$$

$$H^*(H, A) \cong H^*(\Gamma, \mathrm{Coind}_H^\Gamma A). \tag{7}$$

This is Proposition III.6.2 in [10]. Here Ind_H^Γ is the so-called *induced* module $\mathbb{Z}\Gamma \otimes_{\mathbb{Z}H} A$, where $\mathbb{Z}\Gamma$ is considered as a right H-module via the right action of H on Γ, and the left Γ-module structure is given by $g \cdot (g' \otimes a) = gg' \otimes a$ for $g, g' \in \Gamma$, $a \in A$. Similarly, $\mathrm{Coind}_H^\Gamma A$ is the co-induced module $\mathrm{Hom}_{\mathbb{Z}H}(\mathbb{Z}\Gamma, A)$ of H-equivariant maps from the left H-module $\mathbb{Z}\Gamma$ to A. The left action of Γ is defined by

$$(g \cdot f)(g') = f(g'g)$$

for $g, g' \in \Gamma$, $f \in \mathrm{Hom}_{\mathbb{Z}H}(\mathbb{Z}\Gamma, A)$.

Remark 2.1. If the action of H on A is trivial, there is a canonical bijection between $\mathrm{Hom}_{\mathbb{Z}H}(\mathbb{Z}\Gamma, A)$ and $\mathrm{Map}(H\backslash\Gamma, A)$ given by $f \mapsto (Hg \mapsto f(g))$; equipping the latter with the Γ-action $(g \cdot f)(Hg') = f(Hg'g)$ this becomes an isomorphism of Γ-modules. The usual bijection between the sets of left and right cosets given by $Hg \mapsto g^{-1}H$ induces a bijection $\mathrm{Map}(H\backslash\Gamma, A) \to \mathrm{Map}(\Gamma/H, A)$, and the latter also carries a natural left Γ-action making this a Γ-isomorphism, namely $(g \cdot f)(g'H) = f(g^{-1}g'H)$.

We summarize the special case of Shapiro's Lemma we will need in a corollary:

Corollary 2.2. *Let A be an abelian group, and Γ a group which acts transitively on a set R. Consider the Γ-module $\mathrm{Map}(R, A)$ of all maps $R \to A$ with action given by $(g \cdot f)(r) = f(g^{-1}r)$. Let $D \in R$ be any element, and $\Gamma_D \subseteq \Gamma$ the stabilizer subgroup of D. Then there is an isomorphism*

$$H^*(\Gamma, \mathrm{Map}(R, A)) \cong H^*(\Gamma_D, A) \tag{8}$$

where A is considered as a trivial Γ_D-module.

Proof. The bijection $\Gamma/\Gamma_D \to R$ given by $g\Gamma_D \mapsto gD$ clearly induces an isomorphism of Γ-modules $\mathrm{Map}(\Gamma/\Gamma_D, A) \to \mathrm{Map}(R, A)$. Then from Shapiro's Lemma and the isomorphisms mentioned in the above remark we have a sequence of isomorphisms

$$\begin{aligned} H^*(\Gamma_D, A) &\cong H^*(\Gamma, \mathrm{Hom}_{\mathbb{Z}\Gamma_D}(\mathbb{Z}\Gamma, A)) \\ &\cong H^*(\Gamma, \mathrm{Map}(\Gamma_D\backslash\Gamma, A)) \\ &\cong H^*(\Gamma, \mathrm{Map}(\Gamma/\Gamma_D, A)) \\ &\cong H^*(\Gamma, \mathrm{Map}(R, A)). \end{aligned}$$

□

Note that the Γ-module $\mathrm{Map}(R, A)$ can also be considered as the set of all formal A-linear combinations of elements from R (that is, the sum $\sum_{r \in R} m_r r$ corresponds to the map $r \mapsto m_r$).

Specializing to the case $* = 1$, we have

Corollary 2.3. *With assumptions as in Corollary 2.2 above, we have an isomorphism*

$$H^1(\Gamma, \mathrm{Map}(R, A)) \cong H^1(\Gamma_D, A) = \mathrm{Hom}(\Gamma_D, A). \tag{9}$$

We now describe this isomorphism explicitly. First note that a map $u\colon \Gamma \to \mathrm{Map}(R, A)$ can be considered as a map $u\colon \Gamma \times R \to A$ by the adjoint formula $u(g)(r) = u(g, r)$. In this context, the cocycle condition (5) reads

$$u(g_1 g_2, r) = u(g_1, r) + u(g_2, g_1^{-1} r). \tag{10}$$

We wish to derive necessary and sufficient conditions for a cocycle u to be a coboundary δf. For the rest of this section, fix an element $D \in R$ and let $\Gamma_D \subseteq \Gamma$ denote the stabilizer subgroup of D.

Lemma 2.4. *A cocycle $u\colon \Gamma \times R \to A$ is a coboundary if and only if, for every pair $g_1, g_2 \in \Gamma$ with $g_1 g_2^{-1} \in \Gamma_D$, u satisfies the condition*

$$u(g_1, D) = u(g_2, D). \tag{11}$$

Proof. First we prove the necessity of the condition. Suppose that $u = \delta f$ for some $f\colon R \to A$. Since the action is transitive, it is easy to see that the kernel of the map $\delta\colon C^0(\Gamma, \mathrm{Map}(R, A)) \to C^1(\Gamma, \mathrm{Map}(R, A))$ is the set of constant maps $R \to A$. Thus we may without loss of generality assume that $f(D) = 0$. Recall that $u = \delta f$ means that for every $g \in \Gamma$, $r \in R$ we have $u(g, r) = f(r) - f(g^{-1} r)$. In particular,

$$f(g^{-1} D) = -u(g, D). \tag{12}$$

Now if $g_1 g_2^{-1} \in \Gamma_D$, we have $g_1^{-1} D = g_2^{-1} D$, and thus $-u(g_1, D) = f(g_1^{-1} D) = f(g_2^{-1} D) = -u(g_2, D)$ as desired.

Now suppose that u satisfies (11) whenever $g_1 g_2^{-1} D = D$. We need to construct a map $f\colon R \to A$. For $r \in R$, choose $g \in \Gamma$ so that $g^{-1} D = r$, and define f using (12), i.e., $f(r) = f(g^{-1} D) = -u(g, D)$. By assumption, this is a well-defined map (independent of the chosen g), and we only need to check that $u = \delta f$. Let $h \in \Gamma$ and $r \in R$ be arbitrary. To calculate $(\delta f)(h, r)$, we may choose any $g \in \Gamma$ with $g^{-1} D = r$, and we obtain

$$\begin{aligned}(\delta f)(h, r) = f(r) - f(h^{-1} r) &= f(g^{-1} D) - f((gh)^{-1} D) \\ &= -u(g, D) + u(gh, D) = u(h, g^{-1} D) = u(h, r)\end{aligned}$$

by the cocycle condition (10). □

Lemma 2.5. *The restriction of u to $\Gamma_D \times \{D\}$ is a group homomorphism $\tilde{u}\colon \Gamma_D \to A$.*

Proof. Let $g, h \in \Gamma_D$. Then

$$\begin{aligned}\tilde{u}(gh) = u(gh, D) &= u(g, D) + u(h, g^{-1} D) \\ &= u(g, D) + u(h, D) = \tilde{u}(g) + \tilde{u}(h)\end{aligned} \tag{13}$$

as claimed. □

Since A is abelian, $\tilde{u}$ factors through the abelianization $(\Gamma_D)_{\mathrm{ab}}$ of Γ_D, and we have thus established a map

$$\varphi\colon Z^1(\Gamma, \mathrm{Map}(R, A)) \to \mathrm{Hom}(\Gamma_D, A) = \mathrm{Hom}((\Gamma_D)_{\mathrm{ab}}, A).$$

The latter group may be thought of as the cohomology group $H^1((\Gamma_D)_{\mathrm{ab}}, A)$ with trivial action of $(\Gamma_D)_{\mathrm{ab}}$ on A.

Theorem 2.6. *The map φ factors to an isomorphism*

$$H^1(\Gamma, \mathrm{Map}(R, A)) \to H^1((\Gamma_D)_{\mathrm{ab}}, A),$$

which is also denoted φ.

Before we begin the proof, we need an observation: For any cocycle u and any $g \in \Gamma$, $h \in \Gamma_D$ we have

$$\begin{aligned} u(ghg^{-1}, gD) &= u(g, gD) + u(hg^{-1}, D) \\ &= u(g, gD) + u(h, D) + u(g^{-1}, D) \\ &= u(h, D) \end{aligned}$$

using $h^{-1}D = D$ and the fact that $0 = u(1) = u(g^{-1} \cdot g) = u(g^{-1}) + g^{-1}.u(g)$.

Proof of Theorem 2.6. To prove that φ does induce a map on cohomology, we need to show that the restriction of a cobundary δf to $\Gamma_D \times \{D\}$ is identically 0. But this is trivial since

$$\widetilde{\delta f}(h) = (\delta f)(h, D) = f(D) - f(h^{-1}D) = 0$$

for $h \in \Gamma_D$.

Next, assume that the cocycle u restricts to the zero homomorphism $\Gamma_D \to A$. Then for any two elements $g_1, g_2 \in \Gamma$ with $g_1 g_2^{-1} \in \Gamma_D$ we have

$$\begin{aligned} 0 &= u(g_1 g_2^{-1}, D) \\ &= u(g_1, D) + u(g_2^{-1}, g_1^{-1}D) \\ &= u(g_1, D) + u(g_2^{-1})(g_1^{-1}D) \\ &= u(g_1, D) - g_2^{-1}.u(g_2)(g_1^{-1}D) \\ &= u(g_1, D) - u(g_2)(g_2 g_1^{-1}D) \\ &= u(g_1, D) - u(g_2, D) \end{aligned}$$

since $g_2 g_1^{-1} = (g_1 g_2^{-1})^{-1} \in \Gamma_D$, and by Lemma 2.4 we see that u is a coboundary. This shows that φ is injective.

Now, for surjectivity, let $u\colon \Gamma_D \to A$ be any homomorphism. We need to extend u to all of $\Gamma \times R$ in such a way that it becomes a cocycle. To produce this extension, we first assume that an extension exists, and use this to write a formula for a cocycle cohomologous to the given extension. Then we prove that this formula actually defines a cocycle.

Choose a collection $\{h_i\}_{i \in I}$ of representatives for the set $\Gamma_D \backslash \Gamma$ of *right* cosets of Γ_D, and let $1 \in \Gamma$ represent the coset Γ_D. Recall that the map $\Gamma_D \backslash \Gamma \to \Gamma / \Gamma_D$ given by $\Gamma_D x \mapsto x^{-1}\Gamma_D$ is a bijection between the set of right cosets and the set of left cosets of

Γ_D. In particular, $\{h_i^{-1}\}_{i\in I}$ is a collection of representatives of the set of left cosets. We also have a bijection $\Gamma/\Gamma_D \to R$ given by $x\Gamma_D \mapsto xD$.

Now, define $f\colon \Gamma D \to A$ by $f(h_i^{-1}D) = u(h_i, D)$ for $i \in I$. Then since $u(1, D) = 0$, we have $(\delta f)(h_i)(D) = f(D) - f(h_i^{-1}D) = -u(h_i, D)$. Thus, by adding the coboundary of f to the given extension u, we obtain a cohomologous cocycle (again denoted u) satisfying $u(h_i, D) = 0$ for $i \in I$. Furthermore, (the new) u is uniquely determined by its cohomology class and this requirement.

The cocycle condition implies that

$$u(gh_i, D) = u(g, D) + u(h_i, g^{-1}D) = u(g, D) \tag{14}$$

for $i \in I$ and $g \in \Gamma_D$. Since every $x \in \Gamma$ admits a unique factorization as $x = gh_i$ for some $i \in I$ and $g \in \Gamma_D$, this formula extends u to all of $\Gamma \times \{D\}$.

Now consider any $x \in \Gamma$ and $E \in R$. There is a unique $j \in I$ with $h_j^{-1}D = E$, and we have $\Gamma_E = h_j^{-1}\Gamma_D h_j$. Furthermore, the collection $\{h_j^{-1}h_ih_j\}_{i\in I}$ is a collection of representatives for the set $\Gamma_E\backslash\Gamma$ of right cosets of Γ_E. This means that we may factorize x uniquely as $(h_j^{-1}g_0h_j)(h_j^{-1}h_ih_j)$ for some $g_0 \in \Gamma_D$ and $i \in I$. Now we calculate

$$u(x, E) = u(h_j^{-1}g_0h_j \cdot h_j^{-1}h_ih_j, h_j^{-1}D) \tag{15}$$

$$= u(h_j^{-1}g_0h_j, h_j^{-1}D) + u(h_j^{-1}h_ih_j, h_j^{-1}g_0^{-1}h_jh_j^{-1}D) \tag{16}$$

By the observation preceding this proof (with $g = h_j^{-1}$ and $h = g_0$), the first term is equal to the known quantity $u(g_0, D)$. For the second term, we apply the cocycle condition a few more times:

$$\begin{aligned} u(h_j^{-1}h_ih_j, h_j^{-1}D) &= u(h_j^{-1}, h_j^{-1}D) + u(h_ih_j, D) \\ &= -u(h_j, D) + u(h_ih_j, D) \\ &= u(h_ih_j, D) \end{aligned}$$

which is also known since u is known on $\Gamma \times \{D\}$. Thus our formula for the extension of u to all of $\Gamma \times R$ reads

$$u(x, E) = u(g_0, D) + u(h_ih_j, D) \tag{17}$$

where $j \in I$ is the unique index such that $h_j^{-1}D = E$, $i \in I$ is the unique index so that x belongs to the right coset of Γ_E represented by $h_j^{-1}h_ih_j$, and $g_0 = h_jgh_j^{-1}$ is the unique element in Γ_D such that $x = g(h_j^{-1}h_ih_j) = (h_j^{-1}g_0h_j)(h_j^{-1}h_ih_j)$. The second term above is defined by (14); thus one must find the $k \in I$ such that h_ih_j is an element of the right coset of Γ_D represented by h_k, say $h_ih_j = g_1h_k$ for $g_1 \in \Gamma_D$, and then $u(h_ih_j, D) = u(g_1, D)$. It remains to check that (17) defines a cocycle.

Let $x, y \in \Gamma$ and $E \in R$ be arbitrary. As above, there is a unique $j \in I$ with $h_j^{-1}D = E$. Let us try to calculate the right-hand side of the cocycle condition given by $u(xy, E) = u(x, E) + u(y, x^{-1}E)$. We must choose $i \in I$ and $g_1 \in \Gamma_D$ such that

$$x = (h_j^{-1}g_1h_j)(h_j^{-1}h_ih_j) \tag{18}$$

and next we choose $k \in I$ and $g_2 \in \Gamma_D$ such that $h_i h_j = g_2 h_k$. Then

$$u(x, E) = u(g_1, D) + u(g_2, D) = u(g_1 g_2, D)$$

Now, the element $x^{-1}E$ of R is the same as

$$x^{-1}E = h_j^{-1} h_i^{-1} g_1^{-1} h_j E = h_j^{-1} h_i^{-1} D = (h_i h_j)^{-1} D = (g_2 h_k)^{-1} D = h_k^{-1} D$$

so in the calculation of $u(y, x^{-1}E)$ it is h_k which plays the role as h_j in the recipe. This recipe then requires us to find $g_3 \in \Gamma_D$ and $\ell \in I$ such that

$$y = (h_k^{-1} g_3 h_k)(h_k^{-1} h_\ell h_k), \tag{19}$$

and $g_4 \in \Gamma_D$ and $m \in I$ such that $h_\ell h_k = g_4 h_m$. Then

$$u(y, x^{-1}E) = u(g_3, D) + u(g_4, D) = u(g_3 g_4, D).$$

Multiplying x and y using the expressions (18) and (19) and the relations defining the various h'es we obtain

$$\begin{aligned} xy &= (h_j^{-1} g_1 h_i h_j)(h_k^{-1} g_3 h_\ell h_k) \\ &= h_j^{-1} g_1 g_2 g_3 g_4 h_m \end{aligned} \tag{20}$$

On the other hand, the recipe requires us to choose $g \in \Gamma_D$ and $n \in I$ such that

$$xy = h_j^{-1} g h_j h_j^{-1} h_n h_j, \tag{21}$$

and $g' \in \Gamma_D$ and $p \in I$ such that $h_n h_j = g' h_p$. Then $u(xy, E) = u(g, D) + u(g', D)$. Comparing (20) and (21) we see that $g_1 g_2 g_3 g_4 h_m = g h_n h_j$, showing that (by uniqueness of g' and p) $h_p = h_m$ and

$$g' = g^{-1} g_1 g_2 g_3 g_4 \tag{22}$$

Finally we conclude that

$$\begin{aligned} u(xy, E) &= u(g, D) + u(g', D) \\ &= u(g_1 g_2 g_3 g_4, D) \\ &= u(g_1 g_2, D) + u(g_3 g_4, D) \\ &= u(x, E) + u(y, x^{-1}E) \end{aligned}$$

showing that the given recipe in fact defines a cocycle $u\colon \Gamma \times R \to A$. The proof is complete. □

Proof of Theorem 1.1. By the splitting (4), a cocycle $u\colon \Gamma \to \hat{\mathcal{B}}$ is the same as a collection of cocycles $u_D\colon \Gamma \to \hat{M}_D$ for $D \in S$. In fact, thinking of u as a map $\Gamma \times B \to \mathbb{C}$, u_D is simply the restriction of u to $\Gamma \times (\Gamma D)$. Specializing Theorem 2.6 to the case $A = \mathbb{C}$ and $R = \Gamma D$, we see that each u_D is a coboundary if and only if u_D restricted to $\Gamma_D \times \{D\}$ is zero. □

In Section 3 below, we are going to need a theorem linking the low-dimensional cohomology groups of the groups appearing in a short exact sequence. Again quoting from [10] (Corollary VII.6.4)

Theorem 2.7. *Let $1 \to A \to B \to C \to 1$ be a short exact sequence of groups, and M a B-module. Then there is an exact sequence of cohomology groups*

$$0 \to H^1(C, M^A) \to H^1(B, M) \to H^1(A, M)^B \to H^2(C, M^A). \tag{23}$$

Here we regard M as an A-module via restriction of scalars, and then clearly $C \cong B/A$ acts on the submodule M^A invariant under A, making sense of $H^*(C, M^A)$. Since A is normal in B, conjugation defines an action on A by automorphisms, so there is an induced action on cohomology $c_b^*\colon H^*(A, M) \to H^*(A, M)$, where c_b is the map $a \mapsto bab^{-1}$. One may show that A acts trivially on $H^1(A, M)$, so there is an induced action of C, and we have $H^1(A, M)^B = H^1(A, M)^C$.

We need a special case of the above.

Corollary 2.8. *If C is a finite group and $M = \mathbb{C}$ as a trivial B-module, both $H^1(C, \mathbb{C})$ and $H^2(C, \mathbb{C})$ are trivial, so we have an isomorphism*

$$\operatorname{Hom}(B, \mathbb{C}) = H^1(B, \mathbb{C}) \cong H^1(A, \mathbb{C})^B = \operatorname{Hom}(A, \mathbb{C})^B. \tag{24}$$

In other words, the space of homomorphisms $B \to \mathbb{C}$ is isomorphic to the space of homomorphisms $f\colon A \to \mathbb{C}$ satisfying

$$(bf)(a) = f(bab^{-1}) = f(a) \tag{25}$$

for all $a \in A$, $b \in B$.

2.1. Abelianization

If M is a trivial Γ-module, a cocycle $\Gamma \to M$ is the same as a group homomorphism, and all coboundaries vanish identically. Hence, in this situation we have

$$H^1(\Gamma, M) = \operatorname{Hom}(\Gamma, M) = \operatorname{Hom}(\Gamma_{\mathrm{ab}}, M), \tag{26}$$

where $\Gamma_{\mathrm{ab}} = H_1(\Gamma, \mathbb{Z})$ is the abelianization of Γ. In the case where M is torsion free, we further have

$$\operatorname{Hom}(\Gamma_{\mathrm{ab}}, M) = \operatorname{Hom}(\Gamma_{\mathrm{ab}} \otimes \mathbb{Q}, M) = \operatorname{Hom}(H_1(\Gamma, \mathbb{Q}), M). \tag{27}$$

The abelianizations of mapping class groups are all known. Abstractly, they are given as follows:

Theorem 2.9. *Let $\Gamma_{g,r}$ denote the mapping class group of a surface of genus g with r boundary components.*

- *(a) If $g \geq 3$, both $H_1(\Gamma_{g,r}, \mathbb{Z})$ and $H_1(\Gamma_{g,r}, \mathbb{Q})$ are trivial.*
- *(b) For any r, we have $H_1(\Gamma_{2,r}, \mathbb{Z}) \cong \mathbb{Z}/10\mathbb{Z}$, so $H_1(\Gamma_{2,r}, \mathbb{Q}) = 0$.*
- *(c) The group $H_1(\Gamma_{1,0}, \mathbb{Z})$ is cyclic of order 12, whereas for $r \geq 1$, $H_1(\Gamma_{1,r}, \mathbb{Z}) \cong \mathbb{Z}^r$.*
- *(d) For $g = 0$, $H_1(\Gamma_{0,r}, \mathbb{Z})$ is a free abelian group of rank $\binom{r}{2} = r(r-1)/2$, so $H_1(\Gamma_{0,r}, \mathbb{Q})$ is a rational vector space of the same dimension.*

See [11] or [12] for easy proofs of (a), (b) and (c). We will need explicit generators in the genus 1 and 0 cases.

Proposition 2.10. *Consider a torus $\Sigma_{1,r}$ with boundary components $\delta_1, \dots, \delta_r$ for some $r \geq 1$. Let t be a Dehn twist in any non-separating curve, and let d_j denote the Dehn twist along δ_j. Then in $H_1(\Gamma_{1,r}, \mathbb{Z})$, we have the relation*

$$12t = d_1 + \cdots + d_r. \tag{28}$$

The homology classes of $d_1, \dots, d_r$ constitute a basis for $H_1(\Gamma_{1,r}, \mathbb{Q})$.

Proof. For $r \leq 2$, this follows from the chain relation (also known as the two-holed torus relation). On Figure 1, we have obtained $\Sigma_{1,r}$ from $\Sigma_{1,r-1}$ by gluing a pair of pants on to the boundary component δ of $\Sigma_{1,r-1}$ (note that the left-hand part of the surface can contain any number of boundary components). The lantern relation applied to the seven

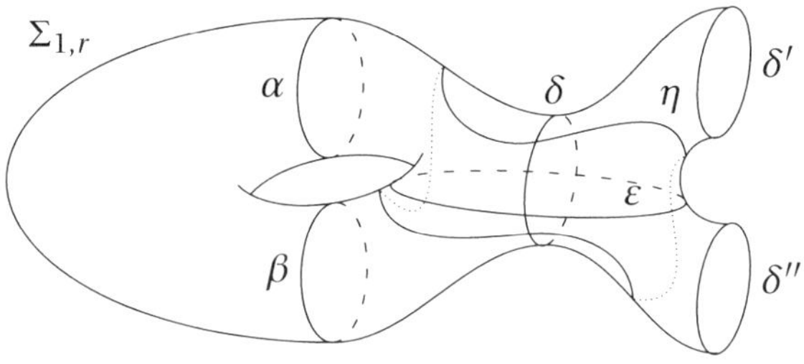

FIGURE 1. Gluing a pair of pants to an $r-1$-holed torus.

curves α, β, δ', δ'', δ, ε, and η yields

$$\tau_\alpha \tau_\beta \tau_{\delta'} \tau_{\delta''} = \tau_\delta \tau_\varepsilon \tau_\eta.$$

Since α, β, ε and η are non-separating in $\Sigma_{1,r}$, the twists in these curves represent the same element in $H_1(\Gamma_{1,r}, \mathbb{Z})$, so we obtain $d' + d'' = d$, where d, d', d'' are the homology classes of τ_δ, $\tau_{\delta'}$ and $\tau_{\delta''}$, respectively.

The last claim follows from the fact that $\Gamma_{1,r}$ is generated by twists in two non-separating curves intersecting in a single point along with the twists in the boundary components. Hence, $H_1(\Gamma_{1,r}, \mathbb{Q})$ is generated by the homology classes of t and $d_1, \dots, d_r$, and (28) shows that t may be omitted. □

It is somewhat more involved to give a symmetric description of $H_1(\Gamma_{0,r}, \mathbb{Z})$. We first recall why this group is free abelian of rank

$$\frac{r(r-1)}{2} = r - 1 + \frac{(r-1)(r-2)}{2}.$$

Let PB_n denote the pure braid group on n strands.

Proposition 2.11. *There is an exact sequence*

$$0 \to \mathbb{Z}^{r-1} \to \Gamma_{0,r} \to \mathrm{PB}_{r-1} \to 1 \tag{29}$$

This sequence splits.

Proof. Think of $\Gamma_{0,r}$ as the mapping class group of a disc with $r-1$ holes. Gluing a disc with a marked point to each of the "inner" boundary components gives a map to the (pure) mapping class group $\Gamma_{0,1}^{r-1}$ of a disc with $r-1$ marked points; the kernel of this map is the free abelian group generated by the twists in the boundary components (see [11], Proposition 3.19), and $\Gamma_{0,1}^{r-1}$ is isomorphic to PB_{r-1} ([11], chapter 9).

Gluing a disc to all of the inner boundary components except the j'th gives a homomorphism from $\Gamma_{0,r}$ to the mapping class group of a cylinder. The latter is infinite cyclic, generated by the Dehn twist in the j'th boundary component. This describes the j'th component of the splitting $\Gamma_{0,r} \to \mathbb{Z}^{r-1}$. □

Since $\Gamma_{0,1}^{r-1}$ is generated by twists τ_{ij} in curves γ_{ij}, $1 \le i < j \le r-1$ encircling the i'th and j'th puncture and all relations are commutation relations (again, see [11], chapter 9), we conclude that $H_1(\Gamma_{0,r}, \mathbb{Z})$ is freely generated by the homology classes of the Dehn twists in $r-1$ of the boundary components, along with Dehn twists along $\binom{r-1}{2} = (r-1)(r-2)/2$ curves separating two of these boundary components from the rest of the surface.

Unfortunately, this way of viewing $H_1(\Gamma_{0,r}, \mathbb{Z})$ singles out one of the boundary components as "special", which makes it inappropriate for our purposes.

Theorem 2.12. *Let A denote the (rank $r + r(r-1)/2$) free abelian group on generators g_i, $1 \le i \le r$ and g_{ij}, $1 \le i < j \le r$. Then $H_1(\Gamma_{0,r}, \mathbb{Z})$ is isomorphic to the quotient of A by the subgroup R generated by the $r+1$ relations*

$$g_k + (r-3)\sum_{i \neq k} g_i = \sum_{\substack{i<j \\ i,j \neq k}} g_{ij} \quad , \quad k = 1, \dots, r \tag{30}$$

$$(r-2)\sum_{i} g_i = \sum_{i<j} g_{ij}. \tag{31}$$

The isomorphism is induced by the map ψ sending g_i to the homology class of the Dehn twist in the i'th boundary component, and g_{ij} to the homology class of the Dehn twist in a curve which bounds a pair of pants together with the i'th and j'th boundary components.

Proof. By the description of $H_1(\Gamma_{0,r}, \mathbb{Z})$ given above, it is clear that ψ is surjective. To prove that ψ descends to a map on the quotient, we must prove that the relations (30) and (31) hold in $H_1(\Gamma_{0,r}, \mathbb{Z})$. The former immediate follows from the generalized lantern relation [13]. Thus, in $H_1(\Gamma_{0,r+1}, \mathbb{Z})$ one has

$$\tau_{r+1} + (r-2) \sum_{i<r+1} \tau_i = \sum_{i<j<r+1} \tau_{ij},$$

and by the epimorphism $\Gamma_{0,r+1} \to \Gamma_{0,r}$ induced by gluing a disc to the last boundary component we see that also (31) holds in $H_1(\Gamma_{0,r}, \mathbb{Z})$.

Since, by the remarks following Proposition 2.11, $H_1(\Gamma_{0,r}, \mathbb{Z})$ is free abelian on the homology classes of the twists τ_i, $1 \leq i < r$ and τ_{ij}, $1 \leq i < j < r$, there is a map $H_1(\Gamma_{0,r}, \mathbb{Z}) \to A$ given by $\tau_i \mapsto g_i$ and $\tau_{ij} \mapsto g_{ij}$. We now prove that the composition of this map with the projection $A \to A/R$ is surjective. This amounts to proving that in A/R, the equivalence classes represented by g_r and g_{kr}, $k < r$, can be written entirely in terms of g_i, $i < r$, and g_{ij}, $1 \leq i < j \leq r - 1$.

Using (30) for $k = r$, this is obvious for g_r. Subtracting (30) from (31), we get

$$(r-3)g_k + \sum_{i \neq k} g_i = \sum_{i<k} g_{ik} + \sum_{k<j<r} g_{kj} + g_{kr},$$

from which we see that each g_{kr} can also be expressed using only $g_1, \ldots, g_{r-1}$ and g_{ij}, $1 \leq i < j < r$ (although g_r occurs on the left-hand side, it can be eliminated by the above). Hence, ψ is invertible. □

3. Computing cohomology

Fix a multicurve D. The purpose of this section is to describe an algorithm to compute $\mathrm{Hom}(\Gamma_D, \mathbb{C})$, the space of homomorphisms from the stabilizer of D to $\mathbb{C}$. We may without loss of generality assume that D does not have components parallel to a boundary component of Σ, since any mapping class preserves each boundary component. We also assume that D is not the empty multicurve (in which case $\Gamma_D = \Gamma$, so $\mathrm{Hom}(\Gamma_D, \mathbb{C})$ is easily computed using Theorem 2.9). We first deal with a simple case.

Lemma 3.1. *If Σ is a closed torus, D consists of parallel copies of a single (non-separating) curve δ. The stabilizer of D is the infinite cyclic group generated by the Dehn twist τ_δ. Hence, $\mathrm{Hom}(\Gamma_D, \mathbb{C})$ is one-dimensional, spanned by the map $\tau_\delta \mapsto 1$.* □

From now on, we will assume that Σ is not a closed torus.

Let d denote the number of distinct isotopy classes of loops occuring in D, and let Δ denote the d-component multicurve consisting of one copy of each of these loops. Clearly Γ_D is a subgroup of Γ_Δ of finite index.

Let Q denote the wreath product $C_2 \wr S_d$ of the cyclic group $C_2 = \{\pm 1\}$ by the symmetric group S_d; in other words, the semi-direct product $C_2^d \rtimes S_d$, where S_d acts on C_2^d by permutation of factors. If we fix an enumeration of the components of Δ and an orientation of each component, we obtain a homomorphism $\Theta\colon \Gamma_\Delta \to Q$ as follows: Write $\Delta = \delta_1 \cup \cdots \cup \delta_d$. Then an element $\varphi \in \Gamma_\Delta$ permutes the components of Δ, so there is $\sigma \in S_d$ with $\varphi(\delta_k) = \delta_{\sigma(k)}$ (as unoriented curves) for all k. The k'th sign is $+1$ if the orientation of $\varphi(\delta_k)$ matches the orientation of $\delta_{\sigma(k)}$, and -1 otherwise. Let K denote the kernel of Θ, i.e., the subgroup of Γ which fixes each component and the orientation of each component, and put $P_D = \Theta(\Gamma_D)$, $P_\Delta = \Theta(\Gamma_\Delta)$. These images depend on the topological types of the multicurves D and Δ.

We let Σ' denote the (possibly non-connected) subsurface of Σ obtained by removing a tubular neighbourhood of Δ. The mapping class group Γ' of Σ' is the direct product of the mapping class groups of the connected components of Σ'. There is a homomorphism $\eta\colon \Gamma' \to \Gamma$ given by extending a diffeomorphism of Σ' by the identity. Theorem 3.18 of [11] yields:

Proposition 3.2. *Assume Σ is not a closed torus. Then the kernel of η is the rank d free abelian group*

$$\ker \eta = \langle \tau_{\alpha_1}\tau_{\beta_1}^{-1}, \dots, \tau_{\alpha_d}\tau_{\beta_d}^{-1} \rangle, \tag{32}$$

where α_k, β_k are the two boundary components of Σ' isotopic (in Σ) to δ_k.

It is clear that the image of η is precisely the group K. The various groups mentioned above are related by the exact sequences

$$\begin{array}{ccccccccc} 1 & \longrightarrow & K & \longrightarrow & \Gamma_D & \xrightarrow{\Theta} & P_D & \longrightarrow & 1 \\ & & \| & & \cap & & \cap & & \\ 1 & \longrightarrow & K & \longrightarrow & \Gamma_\Delta & \xrightarrow{\Theta} & P_\Delta & \longrightarrow & 1 \end{array} \tag{33}$$

and

$$1 \longrightarrow \mathbb{Z}^d \longrightarrow \Gamma' \longrightarrow K \longrightarrow 1. \tag{34}$$

Using these, together with Theorem 2.7 and Corollary 2.8, we see that one may compute $\operatorname{Hom}(\Gamma_\Delta, \mathbb{C})$ as follows:

Proposition 3.3. *As above, let Σ' denote the surface obtained by removing a tubular neighbourhood of Δ. For each boundary component γ of Σ', pick a generator g_γ, and for each pair of boundary components α, β belonging to a genus 0 component of Σ', pick a generator $g_{\alpha,\beta}$. Then $\operatorname{Hom}(\Gamma_\Delta, \mathbb{C})$ is isomorphic to the complex vector space spanned by the g_γ and $g_{\alpha,\beta}$'s, modulo the following relations:*

(a) If γ is a boundary component of Σ' belonging to a component of genus at least 2, then $g_\gamma = 0$.
(b) If γ_1 and γ_2 are boundary components of Σ' arising from the same component of Δ, we have $g_{\gamma_1} = g_{\gamma_2}$.
(c) For each genus 0 component of Σ', the associated generators must satisfy (30) and (31).
(d) If δ_1 and δ_2 are Γ_Δ-related components of Δ (i.e., there is an element of Γ_Δ sending δ_1 to δ_2), the generators arising from the boundary components induced by δ_1 and δ_2 are identified.
(e) If S_1 and S_2 are genus 0 components of Σ', and if there is $\varphi \in \Gamma_\Delta$ with $\varphi(S_1) = S_2$, then for each pair α, β of boundary components of S_1, we have $g_{\alpha,\beta} = g_{\varphi(\alpha),\varphi(\beta)}$.

Proof. A homomorphism from Γ' to $\mathbb{C}$ is the same as a collection of homomorphisms from the mapping class groups of the connected components of Σ' to $\mathbb{C}$. By Theorem 2.9, there are no non-trivial homomorphisms from surfaces of genus 2 or more, and the homomorphisms from the genus 0 and 1 components are exactly described by the generators g_γ,

$g_{\alpha,\beta}$ and the condition (c). Homomorphisms from K to $\mathbb{C}$ are, by (34) and 3.2, in bijection with homomorphisms from Γ' to $\mathbb{C}$ which vanish on elements of the form $\tau_{\alpha_k}\tau_{\beta_k}^{-1}$. Hence, $\mathrm{Hom}(K,\mathbb{C})$ is described by the given generators and relations (a), (b) and (c). Finally, (d) and (e) comes from Corollary 2.8. □

The same algorithm describes $\mathrm{Hom}(\Gamma_D,\mathbb{C})$, except that fewer identifications of generators take place in steps (d) and (e) (since, in addition to the topological type, the multiplicities of the involved components of Δ must match up).

Proof of Theorem 1.2. It is immediate from the isomorphism (2) and the splitting (4) that

$$H^1(\Gamma,\mathcal{O}(\mathcal{M})^*) \cong \prod_D H^1(\Gamma,\hat{M}_D). \tag{35}$$

By Corollary 2.3, we have $H^1(\Gamma,\hat{M}_D) = H^1(\Gamma,\mathrm{Map}(\Gamma D,\mathbb{C})) \cong \mathrm{Hom}(\Gamma_D,\mathbb{C})$. Finally, it is clear by the above algorithm that the latter vector space has finite dimension. □

4. An example

Let Σ be a surface of genus $g \geq 2$ with r boundary components $\sigma_1,\ldots,\sigma_r$. Let Δ be a multicurve consisting of $g-1$ simple closed curves $\delta_1,\ldots,\delta_{g-1}$ such that the complement of (a tubular neighborhood of) Δ is a connected surface Σ' of genus 1. By Theorem 2.9, $H_1(\Gamma',\mathbb{Q})$ is generated by the homology classes of the twists in the $r+2g-2$ boundary components. Hence we see that a basis for $\mathrm{Hom}(K,\mathbb{C})$ is given by the $r+g-1$ elements

$$\begin{aligned} [\tau_{\delta_i}] &\mapsto 1 \qquad i = 1,\ldots,g-1 \\ [\tau_{\sigma_j}] &\mapsto 1 \qquad j = 1,\ldots,r. \end{aligned}$$

Clearly Γ_Δ acts transitively on the set of components of Δ, so in Γ_Δ, all the twists τ_{δ_j} are conjugate. This implies that $\mathrm{Hom}(\Gamma_\Delta,\mathbb{C})$ has dimension $r+1$.

This proves Corollary 1.3 in most cases. In the case of a closed torus, we saw above (Lemma 3.1) that any non-empty multicurve contributes a one-dimensional summand. For a torus with r boundary components, the stabilizer of the empty multicurve (or any multicurve consisting of copies of the boundary components) is the entire mapping class group, which thus contributes an r-dimensional summand by Theorem 2.9.

For a surface of genus 0, the same argument gives a contribution to $H^1(\Gamma,\mathcal{O}(\mathcal{M})^*)$ of dimension $r(r-1)/2$. In fact, for any multicurve D in a surface of genus 0, it is clear that no element of Γ_D may permute the components of D (since every mapping class preserves each of the original boundary components, the connected components of the complement of D cannot be permuted). Hence $K = \Gamma_D$ in the notation above, and we see that the dimension of $\mathrm{Hom}(\Gamma_D,\mathbb{C})$ is given by

$$\sum_S \frac{r_S(r_S-1)}{2} - \#D$$

where the sum runs over the connected components of the complement of (a tubular neighborhood of) D, r_S is the number of boundary components of S, and $\#D$ is the number of components of D.

References

[1] D. Bullock, C. Frohman, and J. Kania-Bartoszyńska, Understanding the Kauffman bracket skein module, *J. Knot Theory Ramifications*, **8** (1999), no. 3, 265–277.

[2] A. R. Skovborg, *The Moduli Space of Flat Connections on a Surface – Poisson Structures and Quantization*, PhD thesis, University of Aarhus, 2006, `http://www.imf.au.dk/publs?id=623`.

[3] S. Gervais, A finite presentation of the mapping class group of a punctured surface, *Topology*, **40** (2001), no. 4, 703–725.

[4] J. E. Andersen, Hitchin's connection, Toeplitz operators and symmetry invariant deformation quantization, to appear in *Quantum Topology*, arXiv preprint, arXiv:math.DG/0611126.

[5] J. E. Andersen, J. Mattes and N. Reshetikhin, The Poisson structure on the moduli space of flat connections and chord diagrams, *Topology*, **35** (1996), no. 4, 1069–1083.

[6] J. E. Andersen, J. Mattes and N. Reshetikhin, Quantization of the algebra of chord diagrams, *Math. Proc. Cambridge Philos. Soc.*, **124** (1998), no. 3, 451–467.

[7] J. E. Andersen and N. L. Gammelgaard, Hitchin's projectively flat connection, Toeplitz operators and the asymptotic expansion of TQFT curve operators, to appear in *Grassmannians, Moduli Spaces and Vector Bundles*, CMI Volume, Editors: David A. Ellwood, Emma Previato and Montserrat Teixidor-i-Bigas (2010), arXiv preprint, arXiv:0903.4091.

[8] J. E. Andersen and R. Villemoes, The first cohomology of the mapping class group with coefficients in algebraic functions on the $SL_2(\mathbb{C})$ moduli space, *Algebr. Geom. Topol.*, **9** (2009), 1177–1199.

[9] J. E. Andersen and R. Villemoes, Cohomology of mapping class groups and the abelian moduli space, to appear in *Quantum Topology*, arXiv preprint, arXiv:0903.4045.

[10] K. S. Brown, *Cohomology of groups*, Graduate Texts in Mathematics, 87, Springer-Verlag, New York, 1982.

[11] B. Farb and D. Margalit, A primer on mapping class groups, January 2011, Version 5.0, available from `http://www.math.utah.edu/~margalit/primer/`.

[12] M. Korkmaz, Low-dimensional homology groups of mapping class groups: a survey, *Turkish J. Math.*, **26** (2002), no. 1, 101–114.

[13] R. W. Bell and D. Margalit, Injections of Artin groups, *Comment. Math. Helv.*, **82** (2007), no. 4, 725–751.

Centre for Quantum Geometry of Moduli Spaces, Ny Munkegade 118, bldg. 1530, Aarhus University, 8000 Aarhus C

E-mail address: `andersen@imf.au.dk`

Centre for Quantum Geometry of Moduli Spaces, Ny Munkegade 118, bldg. 1530, Aarhus University, 8000 Aarhus C

E-mail address: `math@rasmusvillemoes.dk`

Proceedings of 17th Gökova
Geometry-Topology Conference
pp. 79 – 87

Linear ordinary differential equations and Schubert calculus

Boris Shapiro and Michael Shapiro

Abstract. In this short survey we recall some basic results and relations between the qualitative theory of linear ordinary differential equations with real time and the reality problems in Schubert calculus. We formulate a few relevant conjectures.

1. Introduction

Questions asking under what conditions a given enumerative problem in geometry with all real initial data has all real solutions have a long history and appear often in engineering applications. (Below we refer to these as the questions about the total reality of the corresponding enumerative problem.) The most basic and classical of these questions is undoubtedly when a univariate polynomial with real coefficients has all roots real. It goes back to the times of R. Descartes and I. Newton and is very important in connection with the stability problems in control theory developed by (among others) J. C. Maxwell, C. Runge, M. W. Kutta, M. G. Krein.

Another natural test field for questions in total reality is Schubert calculus. In the early 80's W. Fulton revived the interest in these issues by asking whether each enumerative problem in Schubert calculus admits real initial data under which all its solutions will be real. At the same time V. I. Arnold and his school were developing a completely different area, namely, they were studying various generalizations of the classical Sturm theory about the properties of the zeroes of solutions to second order linear ordinary differential equations with real time. It turned out that a natural qualitative theory of linear ordinary differential equations of order greater than 2 is closely related to Schubert calculus.

The purpose of this short note is to review a connection (partially proven and partially conjectural) between the qualitative theory of linear ode of order greater than 2, transversality, and total reality in Schubert calculus.

Key words and phrases. total reality problems, Schubert calculus, osculating flags.
Supported by TÜBİTAK, AIM, and NSF.

2. Sturm theory, disconjugate ODE, and transversality

2.1. Linear ordinary differential equations and curves in vector spaces

Consider a linear homogeneous differential equation (l.o.d.e.) of order n

$$L_n[y] = y^{(n)} + a_1(x)y^{(n-1)} + \ldots + a_n(x)y = 0, \tag{2.1}$$

defined on a given interval I where $a_i(x) \in C^0[I]$.

Denote by V the n-dimensional vector space of solutions of (2.1) considered on I. An arbitrary point $x \in I$ defines the linear evaluation functional on V by assigning to a solution $f \in V$ its value $f(x)$ at x. The image of this evaluation map is a curve $\ell : I \to V^*$ uniquely associated to the original l.o.d.e. Choosing a basis $y_1, \ldots, y_n$ in V we, therefore, identify V^* with $\mathbb{R}^n$. The latter curve will then have the form $\ell(x) = (y_1(x), \ldots, y_n(x)) \in \mathbb{R}^n$ in the standard coordinates. Since the Wronskian determinant does not vanish on I the curve ℓ is non-degenerate, i.e., it has a non-degenerate osculating frame $\{\ell(x), \ell'(x), \ldots, \ell^{(n-1)}(x)\}$ for all $x \in I$.

Note that any solution f of (2.1) is interpreted as a vector in V which, in its turn, defines the corresponding hyperplane $H_f \subset V^*$. The zeros of f correspond to the intersections of ℓ with H_f. The number of zeros of f on I equals the number of intersection points in $\ell \cap H_f$ (counting multiplicities).

For each $1 \leq k \leq n-1$ the curve ℓ induces its *osculating Grassmann curve* $\ell_{G_k(n)} : I \to G_k(n)$ in the Grassmannian $G_k(n)$ of k-dimensional subspaces of $\mathbb{R}^n$. Analogously, one obtains from ℓ its *osculating flag curve* $\ell_{\mathcal{F}} : I \to Fl(n)$ in the variety $Fl(n)$ of complete flags in $\mathbb{R}^n$. Namely, for each point x the corresponding element $\ell_{G_k(n)}(x)$ is the k-dimensional osculating subspace to ℓ at the point x, i.e., the subspace spanned by $\ell(x), \ell'(x), \ldots, \ell^{(k-1)}(x)$. Analogously, the complete flag $\ell_{\mathcal{F}}(x)$ is formed by these subspaces, i.e., it is given by:

$$\left(\operatorname{span}\langle \ell(x)\rangle \subset \operatorname{span}\langle \ell(x), \ell'(x)\rangle \subset \ldots \subset \operatorname{span}\langle \ell(x), \ell'(x), \ldots, \ell^{(n-2)}(x)\rangle \subset \mathbb{R}^n\right).$$

Definition 2.1. Two compete flags $F_\bullet$ and $G_\bullet$ (in the same linear space) are called *transversal* if all their corresponding subspaces are in general position w.r.t. each other, i.e., for all i and j, $1 \leq i, j \leq n$ one has

$$\dim(F_i \cap G_j) = \max(i + j - n, 0),$$

which is the minimal possible value.

Remark 2.1. Flags $F_\bullet$ and $G_\bullet$ are transversal if and only if $\dim(F_i \cap G_{n-i}) = 0$ for all i, $1 \leq i \leq n$.

Definition 2.2. A k-dimensional vector subspace $W^k \subset \mathbb{R}^n$ is transversal to a flag $F_\bullet$ if it is in general position with all subspaces F_r of $F_\bullet$, i.e. $\dim(W^k \cap F_r) = max(k+r-n, 0)$.

Remark 2.2. Clearly, W^k is transversal to $F_\bullet$ if and only if it is transversal to F_{n-k}.

Definition 2.3. The *train* $Tr_{\mathcal{F}}F_\bullet$ (resp., *Grassmann train* $Tr_{G_k(n)}F_\bullet$) of a flag $F_\bullet$ in $\mathbb{R}^n$ is the set of all flags $G_\bullet$ (resp. all k-dimensional subspaces $W^k \in G_k(n)$) such that $G_\bullet$ (resp. W^k) and $F_\bullet$ are not transversal. We can say train in either case when the situation is clear from the context.

Example 2.3. Fix any complete flag $F_\bullet$ in V^* whose hyperplane H_f is dual to the line spanned by a solution f of (2.1). Then all moments x_i of non-transversality of the curve $\ell_{G_1(n)}$ with $F_\bullet$ are exactly the zeros $f(x_i) = 0$. More generally, all moments of non-transversality of the $\ell_{G_k(n)}$ or $\ell_{\mathcal{F}}$ with $F_\bullet$ correspond to intersections of the corresponding osculating Grassmann or flag curve with the corresponding train of flag $F_\bullet$. One can easily identify these moments of non-transversality with the zeros of the Wronskians of k-tuples of solutions of (2.1), see [12].

2.2. Sturm separation theorem

The classical Sturm separation theorem describes the relative position of the roots of two distinct solutions to a linear homogeneous second order differential equation. Namely, the following statement holds.

Theorem 2.1. Let y_1 and y_2 be two non-trivial real solutions of a second order ODE

$$y'' + p(x)y' + q(x)y = 0 \tag{2.2}$$

where $p(x)$ and $q(x)$ are continuous real-valued functions on I that are not multiples of each other. Denote by $\#_1$ (resp. $\#_2$) the number of zeros of y_1 (resp. of y_2) on I. Then between each pair of successive real roots of y_1 there is a root of y_2 and $|\#_1 - \#_2| \leq 1$.

V.I.Arnold generalized the above Sturm theorem to linear Hamiltonian systems with m degrees of freedom having a positive definite time-dependent Hamiltonian (see [1]). The role of zeros in this theory is played by the moments of non-transversality of the Grassmann curve in the Lagrangian Grassmannian to a fixed Lagrangian subspace. (They can also be interpreted as the zeros of certain Wronskians.)

However, at the moment no proven generalization of the Sturm separation theorem in the case of usual higher order l.o.d.e. is known.

Remark 2.4. Notice that by results of Kondratiev [7] for general l.o.d.e. of order greater than 2 no separation theorem can be obtained in terms of zeros of individual solutions.

Note also that we can interpret the zeros of solution f as non-transversality moments between curve $\ell_{G_1(n)}$ and some fixed flag $F_\bullet$ containing hyperplane H_f (see Example 2.3).

Our hope is to obtain a generalization of Sturm separation theorem for higher order l.o.d.e. in terms of the total number of non-transversality moments between $\ell_{\mathcal{F}}$ and some fixed flag $F_\bullet$.

Below we formulate a conjectural generalization of Sturm separation theorem and give some motivation for its validity using the notion of a disconjugate l.o.d.e.

2.3. Disconjugate ODE

Definition 2.4. A l.o.d.e. of order n

$$z^{(n)} + p_1(x)z^{(n-1)} + \ldots + p_n(x)z = 0 \tag{2.3}$$

with real-valued continuous coefficients $p_i(x)$ is called *disconjugate* on an interval I if any of its nontrivial solutions has at most $(n-1)$ zeros on I counting multiplicities.

Example 2.5. The space of solutions of $z^{(n)} = 0$ consists of all polynomials in x of degree less than n, and, therefore, any solution has at most $n-1$ zeroes counting multiplicities on an arbitrary real interval I (as well as on an arbitrary subset of $\mathbb{C}$).

Remark 2.6. Any l.o.d.e. of order n has a nontrivial solution with at least $(n-1)$ zeros. (For example, consider the nontrivial solution of the initial value problem given by $z(x_0) = z'(x_0) = \ldots = z^{(n-2)}(x_0) = 0$, $z^{(n-1)}(x_0) = 1$.) Further, any l.o.d.e. with continuous $p_i(x)$ is locally disconjugate, i.e. for any $x_0 \in I$ there exists its neighborhood such that the above equation is disconjugate in this neighborhood, see e.g. [8].

Recalling the correspondence between nth order l.o.d.e. and non-degenerate curves in $\mathbb{R}^n$ we obtain a geometric interpretation of disconjugacy.

{a l.o.d.e. of order n} $\leftrightarrow$ {a non-degenerate curve in the n-dimensional space V^* (dual to the space of all solutions)}

with

{a disconjugate l.o.d.e.} $\leftrightarrow$ {a space curve in V^* which intersects any hyperplane in V^* at most $n-1$ times (counting multiplicities)}.

Using this correspondence we will call a non-degenerate curve in $\mathbb{R}^n$ *disconjugate* if it intersects any hyperplane in $\mathbb{R}^n$ at most $n-1$ times (counting multiplicities).

The following two lemmas provide criteria of disconjugacy of linear ordinary differential equations (or, equivalently, disconjugate curve) on interval I, compare to [8].

Notation: For a curve $\gamma : I \to \mathbb{R}^n$ we denote by $\gamma_{G_k(n)}$ and $\gamma_{\mathcal{F}}$ the corresponding osculating Grassmann and flag curves, respectively.

Lemma 2.2. (see [12]) A non-degenerate curve γ is *disconjugate* on I if and only if for all $t_1 \neq t_2 \in I$ one has that $\gamma_{\mathcal{F}}(t_1)$ is transverse to $\gamma_{\mathcal{F}}(t_2)$.

Lemma 2.3. (see [12]) A non-degenerate curve γ is *conjugate* (i.e. **not disconjugate**) if and only if for any complete flag $G_\bullet$ there is some $t \in I$ such that $G_\bullet$ and $\gamma_{\mathcal{F}}(t)$ are non-transversal.

2.4. Conjectural multiplicative Sturm separation theorem

One can split the time interval I for any equation (2.1) into maximally disconjugate subintervals. Instead of individual solutions one should compare different fundamental solutions, i.e., count the number of moments of non-transverality of the flag curve to the trains of two different complete flags. Lemmata 2.3 and 2.2 then give some estimate of the number of non-transversalities on each of the disconjugate subintervals. This idea leads to the following conjecture which is a generalization of Sturm separation theorem to the case of higher order l.o.d.e.'s.

For a non-degenerate curve $\gamma : I \to \mathbb{R}^n$ and any pair of fixed flags $G_\bullet$ and $\hat{G}_\bullet$ we denote by $\#_{\mathcal{F}}$ (resp. $\widehat{\#_{\mathcal{F}}}$) the number of moments of non-transversality between $\gamma_{\mathcal{F}}(t)$ and $G_\bullet$ (resp. $\widehat{G_\bullet}$).

Conjecture 2.4. Let γ be a non-degenerate curve in $\mathbb{R}^n$, $n \geq 2$. Then

$$\widehat{\#_{\mathcal{F}}} \leq \frac{n^3 - n}{6} \left(\#_{\mathcal{F}} + 1\right). \tag{2.4}$$

Remark 2.7. Note that $\frac{n^3-n}{6} = \sum_k \dim G_k(n)$.

Kondratiev's results show that one can not hope to get similar estimates to (2.4) in terms of nontransversality moments for individual Grassmannians $G_k(n)$. However, Conjecture 2.4 would follow from Conjecture 2.5 below.

Definition 2.5. A curve in $G_k(n)$ is called *Grassmann convex* if it intersects the train of any flag at most $\dim G_k(n) = k(n-k)$ times.

Conjecture 2.5. If γ is a non-degenerate disconjugate curve in $\mathbb{R}^n$ then its osculating Grassmann curve $\gamma_{G_k(n)} \subset G_k(n)$ is Grassman convex.

Remark 2.8. Conjecture 2.5 is evident in $\mathbb{R}^3$, proven in $\mathbb{R}^4$ see [16], and open in $\mathbb{R}^n$ for $n \geq 5$.

2.5. Local geometry of osculating flag curves

Examples in low dimensions led us to the following conjectures on the local geometry of osculating flag curves.

Recall that the train $TrF_\bullet$ of any flag $F_\bullet$ is an algebraic hypersurface in the space of (complete) flags. If $x \in TrF_\bullet$ then $TrF_\bullet$ separates a sufficiently small open ball B in the space of flags centered at x into a finite number of connected components.

Conjecture 2.6. Let γ be the germ of a non-degenerate curve in $\mathbb{R}^n$ and $\gamma_{\mathcal{F}}(0) \in TrF_\bullet$. Then the germ of the osculating flag curve $\gamma_{\mathcal{F}}$ crosses $TrF_\bullet$ and goes from one connected component of $B \setminus TrF_\bullet$ to another one, i.e., for a sufficiently small $\delta > 0$ flags $\gamma_{\mathcal{F}}(-\delta)$ and $\gamma_{\mathcal{F}}(\delta)$ belong to different connected components of $B \setminus TrF_\bullet$.

Conjecture 2.7. If the osculating curve $\gamma_{\mathcal{F}}$ of a disconjugate non-degenerate curve γ passes from any connected component C of $B \setminus TrF_\bullet$ to another component then it never returns back to C.

Remark 2.9. Conjectures 2.6 and 2.7 would imply a weaker version of Conjecture 2.4.

Conjecture 2.8. Let γ be a non-degenerate curve in $\mathbb{R}^n$, $n \geq 2$. Then, there is a positive integer constant $K(n)$ depending on n only such that

$$\widehat{\#_{\mathcal{F}}} \leq K(n)\,(\#_{\mathcal{F}}+1)\,. \tag{2.5}$$

3. Transversality and M-varieties

In this section we discuss a relation between transversality and the so-called M-property of intersections of Schubert varieties with the emphasis on intersections of trains of osculating flags to a non-degenerate curve γ. For the osculating flag $\gamma_{\mathcal{F}}(t)$ we call the time moment t its *reference point.*

Note first that (the proof) of Lemma 2.2 implies that the trains $Tr\gamma_{\mathcal{F}}(t_1)$ and $Tr\gamma_{\mathcal{F}}(t_2)$ of any two distinct osculating flags to a disconjugate curve γ are transversal. Therefore, the topology of their intersection does not change when we change reference points t_1 and t_2.

Unfortunately, as it was shown in [10] for general disconjugate curves such transversality fails for intersections of the trains of more than two osculating flags.

However, the so-called dimensional transversality holds for intersection of Grassmann trains of osculating flags to the moment curve. The following result was proved by Eisenbud and Harris.

Lemma 3.1. (*Dimensional transversality,* see [6]) Suppose that $t_1 < t_2 < \ldots < t_r$ be the set of reference points on the moment curve ν in $\mathbb{R}^n$. Let $Sch_{\gamma_{\mathcal{F}}(t_i)}$ be the Schubert decomposition of Grassmannian $G_k(n)$with respect to the osculating flag $\nu_{\mathcal{F}}(t_i)$. Then flags $\nu_{\mathcal{F}}(t_i)$ have *dimensional transversality property,* i.e. the codimension of the intersection of an arbitrary set of cells $C_1, \ldots, C_r$, where C_i belongs to $Sch_{\nu_{\mathcal{F}}(t_i)}$, equals the maximum between $dim(G_k(n))$ and the sum of codimensions of C_is.

It implies, in particular, that intersections of $k(n-k)$-tuple of Grassmann trains in $G_k(n)$ of osculating flags to the moment curve is pure zero-dimensional, i.e., contains points only. Moreover, any intersection of more trains is empty.

The stronger transversality statement (see Conjecture 3.2) might be the actual reason why the total reality property holds for enumerative properties in Schubert calculus in Grassmannians (see [18]).

Below we discuss the total reality and the M-property of such problems. Let us first recall the notion of an M-variety.

Example 3.1. ***M*-curves.** Let $C = C^{\mathbb{C}}$ be a real (i.e., invariant under the standard complex conjugation) projective nonsingular algebraic curve of genus g embedded in $\mathbb{C}P^n$, and $C^{\mathbb{R}}$ be its real part, i.e., $C^{\mathbb{R}} = C^{\mathbb{C}} \cap \mathbb{R}P^n$. Then Harnack-Klein theorem claims that the number of connected components of $C^{\mathbb{R}}$ does not exceed $g+1$. The curve C is called an *M-curve* if the latter number of connected components *equals* $g+1$.

Furthermore, for any real (invariant under the complex conjugation) embedded projective algebraic variety $X = X^{\mathbb{C}} \subset \mathbb{C}P^n$, and its real part $X^{\mathbb{R}} = X \cap \mathbb{R}P^n$ the

Smith inequality claims that:

$$\sum b_i(X^{\mathbb{R}}, \mathbb{Z}_2) \leq \sum b_i(X^{\mathbb{C}}, \mathbb{Z}_2). \tag{3.1}$$

We say that X is an *M-variety* if $\sum b_i(X^{\mathbb{R}}, \mathbb{Z}_2) = \sum b_i(X^{\mathbb{C}}, \mathbb{Z}_2)$. (Roughly speaking, $X = X^{\mathbb{C}}$ is an M-variety if any cycle in $X^{\mathbb{C}}$ generates a cycle in $X^{\mathbb{R}}$.)

Example 3.2. The intersection of two open Schubert cells in general position in the space of complete flags is an M-variety [17].

Example 3.3. The same holds for intersections of two or more open real Schubert cells in the space of special incomplete two step flags consisting of a line in a hyperplane, see [15].

Conjecture 3.2. For the moment curve $\nu(t) = (1 : t : \ldots : t^{n-1})$ and any set of reference points $t_1 < t_2 < \ldots < t_m$ the corresponding Grassmann trains $Tr\nu_{G_k(n)}(t_i)$, $1 \leq i \leq m$ intersect transversally in $G_k(n)$.

F. Sottile in [19] proved that the special Schubert calculus is *totally real*. More exactly, he showed that there exist reference points

$$t_1 < t_2 < t_3 < \ldots < t_{k(n-k)} \tag{3.2}$$

on the moment curve ν in $\mathbb{R}^n$ such that all intersection points $\cap_{j=1}^{k(n-k)} Tr\nu_{G_k(n)}(t_j)$ are real.

Remark 3.4. Note that if a variety $X \subset \mathbb{C}P^n$ of pure dimension zero contains only real points then it is an M-variety.

The idea of transversality combined with the total reality of the special Schubert calculus leads to the following

Conjecture 3.3. (Total reality conjecture) Let $t_1 < \ldots < t_{k(n-k)}$ be an arbitrary $k(n-k)$-tuple of distinct reference points on the moment curve ν. Then the intersection $\cap_{j=1}^{k(n-k)} Tr\nu_{G_k(n)}(t_j)$ in any Grassmann manifold $G_k(n)$ is an M-variety.

Indeed, transversality means that all such intersections have the same topology for any positions of reference points. In particular, for the dimension zero the number of points would be the same.

The general transversality conjecture 3.2 remains widely open. But its zero-dimensional version and, in particular, conjecture 3.3 is proved. Conjecture 3.3 for $G_{n-2}(n)$ (or, by duality for $G_2(n)$) was proved by A. Eremenko and A. Gabrielov in [4]. The general case of $G_k(n)$ was proved by A. Varchenko, E. Mukhin, and V. Tarasov see [9].

As we mentioned above it was shown in [10] that if we replace the moment curve by a general disconjugate curve the analog of conjecture 3.3 fails. On the other hand, it is clear, that the set of curves for which conjecture 3.3 is valid is open in the C^n-topology on the space of smooth nondegenerate curves in $\mathbb{R}^n$.

Question: *Find an open neighborhood of the moment curve such that each curve in this neighborhood satisfies* total reality conjecture 3.3.

4. Total reality of meromorphic functions

In [4] the authors settled the total reality conjecture for $G_{n-2}(n)$ using its equivalent reformulation given below.

Namely, let $f = \dfrac{p(t)}{q(t)} = \dfrac{\sum_{i=0}^{n} p_i t^i}{\sum_{i=0}^{n} q_i t^i}$ be a rational function of degree n. Such a rational function determines a subspace $S_f \subset \mathbb{R}^{n+1}$ of codimension 2 given as follows. If $\{x_0, x_1, \dots, x_n\}$ are standard coordinates in $\mathbb{R}^{n+1}$ then S_f is the intersection of two hyperplanes given by the equations $\sum_{i=0}^{n} p_i x_i = 0$ and $\sum_{i=0}^{n} q_i x_i = 0$.

Remark 4.1. If $\det \begin{pmatrix} a & b \\ c & d \end{pmatrix} \neq 0$, $a, b, c, d \in \mathbb{C}$ then we call functions $f(t)$ and $\dfrac{af(t)+b}{cf(t)+d}$ *Möbius-equivalent.*

Remark 4.2. Möbius-equivalent functions determine the same codimension 2 subspace.

Direct computation shows that the dimension two osculating subspace to the moment curve ν at the reference point t_0 intersects S_f if and only if $f'(t_0) = 0$.

Then the total reality conjecture can be formulated in the following way.

Equivalent reformulation: *Let* $f = \dfrac{p(t)}{q(t)}$ *be a rational function of degree* n *such that all its* $(2n-2)$ *critical points are real and distinct. Then* f *is Möbius-equivalent to a real function* g, *i.e., all coefficients of numerator and denominator of* g *are real.*

Inspired by the results of [4] for rational functions we want to ask whether total reality might hold for meromorphic functions on a real algebraic curve of a positive genus.

Question: *If* Σ *is a real curve of a positive genus and* ϕ *is a meromorphic function on* Σ *with all its critical points real, is it true that* ϕ *is Möbius-equivalent to a real meromorphic function?*

In conclusion, let us mention two partial cases when the above question has a positive answer.

Theorem 4.1. (See [5]) *A function of any prime degree* d *on any real curve of genus* $g > \frac{d^2-4d+3}{3}$ *whose critical points are real and distinct, is Möbius-equivalent to a real meromorphic function.*

Theorem 4.2. (See [3]) *Any function of degree at most four on any real curve whose critical points are real and distinct is Möbius-equivalent to a real meromorphic function.*

Acknowledgements. The authors want to acknowledge the hospitality of AIM in Palo Alto during the program 'Algebraic systems with only real solutions' held in October 2010 where this material was presented. The second author expresses his gratitude to TUBITAK and NSF for support and to organizers of the Gökova 2010 conference for hospitality.

References

[1] V. I. Arnold, Sturm theorems and symplectic geometry, (Russian), *Funktsional. Anal. i Prilozhen*, **19** (1985), no. 4, 1–10, 95.

[2] A. Berenstein, S. Fomin and A. Zelevinski, Parametrizations of canonical bases and totally positive matrices, *Adv. Math.* **122** (1996), 49–149.

[3] A. Degtyarev, T. Ekedahl, I. Itengberg, B. Shapiro and M. Shapiro, On total reality of meromorphic functions, *Ann. Inst. Fourier (Grenoble)*, **57** (2007), no. 6, 2015–2030.

[4] A. Eremenko and A. Gabrielov, Rational functions with real critical points and the B. and M. Shapiro conjecture in real enumerative geometry, *Ann. of Math. (2)*, **155** (2002), no. 1, 105–129.

[5] T. Ekedahl, B. Shapiro and M. Shapiro, First step towards total reality of meromorphic functions, *Mosc. Math. J.* **6** (2006), no. 1, 95–106.

[6] D. Eisenbud and J. Harris, Divisors on general curves and cuspidal rational curves, *Inv.Math.* **74** (1983), 374–418.

[7] V. A. Kondratiev, Oscillatory properties of solutions of the equation $y^{(n)} + p(y) = 0$, *Trudy Moskovskogo Matematicheskogo Obshestva* **10** (1961), 419–436.

[8] A. Ju. Levin, Disconjugacy of solutions of equations $x^{(n)} + p_1(t)x^{(n-1)} + ... + p_n(t)x = 0$, *Soviet Math. Surveys*, **24** (1969), 43–96.

[9] E. Mukhin, V. Tarasov and A. Varchenko, The B. and M. Shapiro conjecture in real algebraic geometry and the Bethe ansatz, to appear in *Ann. of Math.*, arXiv preprint, arXiv:math/0512299v2

[10] V. Sedykh and B. Shapiro, On two conjectures concerning convex curves, *Internat. J. Math.* **16** (2005), no. 10, 1157–1173.

[11] B. Shapiro and M. Shapiro, On the boundary of totally positive upper triangular matrices, *Linear Algebra Appl.* **231** (1995), 105–109.

[12] B. Shapiro, Spaces of linear differential equations and flag manifolds. (Russian) *Izv. Akad. Nauk SSSR Ser. Mat.* **54** (1990), no. 1, 173–187 (English transl. in *Math. USSR - Izv.* **36** (1991), no. 1, 183–197.

[13] B. Shapiro, M. Shapiro and A. Vainshtein, Connected components in the intersection of two open opposite Schubert cells in $SL_n(R)/B$, *Internat. Math. Res. Notices* (1997), no. 10, 469–493.

[14] B. Shapiro, M. Shapiro and A. Vainshtein, Skew-symmetric vanishing lattices and intersection of Schubert cells, *Internat. Math. Res. Notices* (1998), no. 11, 563–588.

[15] B. Shapiro and M. Shapiro, The M-property of flag varieties, *Topology Appl.* **43** (1992), no. 1, 65–81.

[16] B. Shapiro and M. Shapiro, Projective convexity in $\mathbb{P}^3$ implies Grassmann convexity, *Internat. J. Math.* **11** (2000), no. 4, 579–588.

[17] M. Shapiro, *Nonoscillating differential equations*, Ph.D. Thesis, (1992), Moscow State University.

[18] F. Sottile, Frontiers of reality in Schubert Calculus, *Bull. AMS*, **47** (2010), no. 1, 31–71.

[19] F. Sottile, Special Schubert calculus is real, *Electronic Res. Announcements*, AMS, **5** (1999), 35–39.

Department of Mathematics, Stockholm University, SE-106 91 Stockholm, Sweden
E-mail address: shapiro@math.su.se

Department of Mathematics, Michigan State University, East Lansing, MI 48824-1027
E-mail address: mshapiro@math.msu.edu

Proceedings of 17th Gökova
Geometry-Topology Conference
pp. 88 – 109

Alexandrov meets Kirszbraun

Stephanie Alexander, Vitali Kapovitch and Anton Petrunin

Abstract. We give a simplified proof of the generalized Kirszbraun theorem for Alexandrov spaces, which is due to Lang and Schroeder. We also discuss related questions, both solved and open.

1. Introduction

Kirszbraun's theorem states that any *short map* (i.e., 1-Lipschitz map) from a subset of Euclidean space to another in Euclidean space can be extended as a short map to the whole space.

This theorem was proved first by Kirszbraun in [6]. Later it was reproved by Valentine in [14] and [15], where he also generalized it to pairs of Hilbert spaces of arbitrary dimension as well as pairs of spheres of the same dimension and pairs of hyperbolic spaces with the same curvature. J. Isbel in [5] studied target spaces that satisfy the above condition for any source space.

Valentine was also interested in pairs of metric spaces, say $\mathcal{U}$ and $\mathcal{L}$, which satisfy the above property, namely, given a subset $Q \subset \mathcal{U}$, any short map $Q \to \mathcal{L}$ can be extended to a short map $\mathcal{U} \to \mathcal{L}$. It turns out that this property has a lot in common with the definition of Alexandrov spaces (see theorems 5.1, 3.1 and 3.2). Surprisingly, this relationship was first discovered only in the 1990's; it was first published by Lang and Schroeder in [9]. (The third author of this paper came to similar conclusions a couple of years earlier, and told it to the first author, but did not publish the result.)

We slightly improve the results of Lang and Schroeder. Our proof is based on the barycentric maps used by Kleiner in [7]. These maps originate from earlier work on Riemannian center of mass, as introduced by Grove and Karcher [4]. As far as we know Kleiner was the first to focus on the importance of the barycentric maps considered here.

The material of this paper will be included in the book on Alexandrov geometry that we currently are writing. We hope our paper will spur further progress on the questions raised here, thereby improving the book. In any case, since the book will not appear for several years, we would like to let people know about the present state of the art around Kirszbraun's Theorem.

Structure of the paper. We introduce notations in Section 2. In section 3 we give alternative definitions of Alexandrov spaces based on the Kirszbraun property for 4-point sets. The generalized Kirszbraun theorem is proved in Section 5. In the sections

4 and 6 we describe some comparison properties of finite subsets of Alexandrov spaces. In Section 7 we discuss related open problems. Appendices A and B describe Kleiner's barycentric map and an analog of Helly's theorem for Alexandrov spaces.

Historical remark. Not much is known about the author of this remarkable theorem. The theorem appears in Kirszbraun's master's thesis which he defended in Warsaw University in 1930. His name is Mojżesz and his second name is likely to be Dawid but is uncertain. He was born either in 1903 or 1904 and died in a ghetto in 1942. After university, he worked as an actuary in an insurance company; [6] seems to be his only publication in mathematics.

Acknowledgment. We want to thank S. Ivanov, N. Lebedeva and A. Lytchak for useful comments and pointing out misprints. Also we want to thank L. Grabowski for bringing to our attention the entry about Kirszbraun in the Polish Biographical Dictionary.

2. Preliminaries

In this section we mainly introduce our notations.

Metric spaces. Let $\mathcal{X}$ be a metric space. The distance between two points $x, y \in \mathcal{X}$ will be denoted as $|xy|$ or $|xy|_{\mathcal{X}}$.

Given $R \in [0, \infty]$ and $x \in \mathcal{X}$, the sets

$$\mathrm{B}(x, R) = \{y \in \mathcal{X} \mid |xy| < R\},$$
$$\overline{\mathrm{B}}[x, R] = \{y \in \mathcal{X} \mid |xy| \leqslant R\}.$$

are called respectively the *open* and *closed ball* of radius R with center at x.

A metric space $\mathcal{X}$ is called *intrinsic* if for any $\varepsilon > 0$ and any two points $x, y \in \mathcal{X}$ with $|xy| < \infty$ there is an ε-midpoint for x and y; i.e., there is a point $z \in \mathcal{X}$ such that $|xz|, |zy| < \frac{1}{2}\cdot|xy| + \varepsilon$.

Model space. $\mathbb{M}^m[\kappa]$ denotes m-dimensional model space with curvature κ; i.e., the simply connected m-dimensional Riemannian manifold with constant sectional curvature κ.

Set $\varpi^\kappa = \operatorname{diam} \mathbb{M}^2[\kappa]$, so $\varpi^\kappa = \infty$ if $\kappa \leqslant 0$ and $\varpi^\kappa = \pi/\sqrt{\kappa}$ if $\kappa > 0$. (The letter ϖ is a glyph variant of lower case π, but is usually pronounced as *pomega*.)

Ghost of Euclid. Let $\mathcal{X}$ be a metric space and $\mathbb{I}$ be a real interval. A globally isometric map $\gamma\colon \mathbb{I} \to \mathcal{X}$ will be called a *unitspeed geodesic*. A unitspeed geodesic between p and q will be denoted by $\mathrm{geod}_{[pq]}$. We consider $\mathrm{geod}_{[pq]}$ with parametrization starting at p; i.e., $\mathrm{geod}_{[pq]}(0) = p$ and $\mathrm{geod}_{[pq]}(|pq|) = q$. The image of $\mathrm{geod}_{[pq]}$ will be denoted by $[pq]$ and called a *geodesic*.

Also we will use the following short-cut notation:

$$]pq[= [pq]\backslash\{p, q\}, \qquad]pq] = [pq]\backslash\{p\}, \qquad [pq[= [pq]\backslash\{q\}.$$

A metric space $\mathcal{X}$ is called *geodesic* if for any two points $x, y \in \mathcal{X}$ there is a geodesic $[xy]$ in $\mathcal{X}$.

Given a geodesic $[pq]$, we denote by $\operatorname{dir}[pq]$ its direction at p. We may think of $\operatorname{dir}[pq]$ as belonging to the space of directions Σ_p at p, which in turn can be identified with the unit sphere in the tangent space T_p at p. Further we set $\log[pq] = |pq|\cdot\operatorname{dir}[pq]$; it is a *tangent vector* at p, that is, an element of T_p.

For a triple of points $p, q, r \in \mathcal{X}$, a choice of triple of geodesics $([qr], [rp], [pq])$ will be called a *triangle* and we will use the notation $[pqr] = ([qr], [rp], [pq])$. If p is distinct from x and y, a pair of geodesics $([px], [py])$ will be called a *hinge*, and denoted by $[p\,{}^x_y] = ([px], [py])$.

Functions. A locally Lipschitz function f on a metric space $\mathcal{X}$ is called λ-convex (λ-concave) if for any geodesic $\operatorname{geod}_{[pq]}$ in $\mathcal{X}$ the real-to-real function

$$t \mapsto f \circ \operatorname{geod}_{[pq]}(t) - \tfrac{\lambda}{2}\cdot t^2$$

is convex (respectively concave). In this case we write $f'' \geqslant \lambda$ (respectively $f'' \leqslant \lambda$).

A function f is called *strongly convex* (*strongly concave*) if $f'' \geqslant \delta$ (respectively $f'' \leqslant -\delta$) for some $\delta > 0$.

Model angles and triangles. Let $\mathcal{X}$ be a metric space, $p, q, r \in \mathcal{X}$ and $\kappa \in \mathbb{R}$. Let us define a *model triangle* $[\tilde p\tilde q\tilde r]$ (briefly, $[\tilde p\tilde q\tilde r] = \tilde\triangle^\kappa(pqr)$) to be a triangle in the model plane $\mathbb{M}^2[\kappa]$ such that

$$|\tilde p\tilde q| = |pq|, \quad |\tilde q\tilde r| = |qr|, \quad |\tilde r\tilde p| = |rp|.$$

If $\kappa \leqslant 0$, the model triangle is said to be defined, since such a triangle always exists and is unique up to an isometry of $\mathbb{M}^2[\kappa]$. If $\kappa > 0$, the model triangle is said to be defined if in addition

$$|pq| + |qr| + |rp| < 2\cdot\varpi^\kappa.$$

In this case the triangle also exists and is unique up to an isometry of $\mathbb{M}^2[\kappa]$.

If for $p, q, r \in \mathcal{X}$, the model triangle $[\tilde p\tilde q\tilde r] = \tilde\triangle^\kappa(pqr)$ is defined and $|pq|, |pr| > 0$, then the angle measure of $[\tilde p\tilde q\tilde r]$ at $\tilde p$ will be called the *model angle* of the triple p, q, r, and will be denoted by $\tilde\measuredangle^\kappa(p\,{}^q_r)$.

Curvature bounded below. We will denote by $\overline{\mathrm{CBB}}[\kappa]$, complete intrinsic spaces $\mathcal{L}$ with curvature $\geqslant \kappa$ in the sense of Alexandrov. Specifically, $\mathcal{L} \in \overline{\mathrm{CBB}}[\kappa]$ if for any quadruple of points $p, x^1, x^2, x^3 \in \mathcal{U}$, we have

$$\tilde\measuredangle^\kappa(p\,{}^{x^1}_{x^2}) + \tilde\measuredangle^\kappa(p\,{}^{x^2}_{x^3}) + \tilde\measuredangle^\kappa(p\,{}^{x^3}_{x^1}) \leqslant 2\cdot\pi. \qquad ❶$$

or at least one of the model angles $\tilde\measuredangle^\kappa(p\,{}^{x^i}_{x^j})$ is not defined.

Condition ❶ will be called *(1+3)-point comparison.*

According to Plaut's theorem [11, Th. 27], any space $\mathcal{L} \in \overline{\mathrm{CBB}}$ is G_δ-geodesic; that is, for any point $p \in \mathcal{L}$ there is a dense G_δ-set $W_p \subset \mathcal{L}$ such that for any $q \in W_p$ there is a geodesic $[pq]$.

We will use two more equivalent definitions of $\overline{\mathrm{CBB}}$ spaces (see [1]). Namely, a complete G_δ-geodesic space is in $\overline{\mathrm{CBB}}$ if and only if it satisfies either of following conditions:

(1) (point-on-side comparison) For any geodesic $[xy]$ and $z \in]xy[$, we have

$$\tilde{\measuredangle}^\kappa(x\,{}^p_y) \leqslant \tilde{\measuredangle}^\kappa(x\,{}^p_z); \qquad ❷$$

or, equivalently,

$$|\tilde{p}\tilde{z}| \leqslant |pz|,$$

where $[\tilde{p}\tilde{x}\tilde{y}] = \tilde{\triangle}^\kappa(pxy)$, $\tilde{z} \in]\tilde{x}\tilde{y}[$, $|\tilde{x}\tilde{z}| = |xz|$.

(2) (hinge comparison) For any hinge $[x\,{}^p_y]$, the angle $\measuredangle[x\,{}^p_y]$ is defined and

$$\measuredangle[x\,{}^p_y] \geqslant \tilde{\measuredangle}^\kappa(x\,{}^p_y).$$

Moreover, if $z \in]xy[$, $z \neq p$ then for any two hinges $[z\,{}^p_y]$ and $[z\,{}^p_x]$ with common side $[zp]$

$$\measuredangle[z\,{}^p_y] + \measuredangle[z\,{}^p_x] \leqslant \pi.$$

We also use the following standard result in Alexandrov geometry, which follows from the discussion in the survey of Plaut [11, 8.2].

2.1. Theorem. *Let $\mathcal{L} \in \overline{\mathrm{CBB}}$. Given an array of points $(x^1, x^2 \ldots, x^n)$ in $\mathcal{L}$, there is a dense G_δ-set $W \subset \mathcal{L}$ such that for any $p \in W$, all the directions $\mathrm{dir}[px^i]$ lie in an isometric copy of a unit sphere in Σ_p. (Or, equivaletntly, all the vectors $\log[px^i]$ lie in a subcone of the tangent space T_p which is isometric to Euclidean space.)*

Curvature bounded above. We will denote by $\mathrm{CAT}[\kappa]$ the class of metric spaces $\mathcal{U}$ in which any two points at distance $< \varpi^\kappa$ are joined by a geodesic, and which have curvature $\leqslant \kappa$ in the following global sense of Alexandrov: namely, for any quadruple of points $p^1, p^2, x^1, x^2 \in \mathcal{U}$, we have

$$\tilde{\measuredangle}^\kappa(p^1\,{}^{x^1}_{x^2}) \leqslant \tilde{\measuredangle}^\kappa(p^1\,{}^{p^2}_{x^1}) + \tilde{\measuredangle}^\kappa(p^1\,{}^{p^2}_{x^2}), \text{ or } \tilde{\measuredangle}^\kappa(p^2\,{}^{x^1}_{x^2}) \leqslant \tilde{\measuredangle}^\kappa(p^2\,{}^{p^1}_{x^1}) + \tilde{\measuredangle}^\kappa(p^2\,{}^{p^1}_{x^2}), \qquad ❸$$

or one of the six model angles above is undefined.

The condition ❸ will be called *(2+2)-point comparison* (or *(2+2)-point κ-comparison* if a confusion may arise).

We denote the complete $\mathrm{CAT}[\kappa]$ spaces by $\overline{\mathrm{CAT}}[\kappa]$.

The following lemma is a direct consequence of the definition:

2.2. Lemma. *Any complete intrinsic space $\mathcal{U}$ in which every quadruple p^1, p^2, x^1, x^2 satisfies the (2+2)-point κ-comparison is a $\overline{\mathrm{CAT}}[\kappa]$ space (that is, any two points at distance $< \varpi^\kappa$ are joined by a geodesic).*

In particular, the completion of a $\mathrm{CAT}[\kappa]$ space again lies in $\mathrm{CAT}[\kappa]$.

We have the following basic facts (see [1]):

2.3. Lemma. *In a $\mathrm{CAT}[\kappa]$ space, geodesics of length $< \varpi^\kappa$ are uniquely determined by, and continuously dependent on, their endpoint pairs.*

2.4. Lemma. *In a $\mathrm{CAT}[\kappa]$ space, any open ball $\mathrm{B}(x, R)$ of radius $R \leqslant \varpi^\kappa/2$ is convex, that is, $\mathrm{B}(x, R)$ contains every geodesic whose endpoints it contains.*

We also use an equivalent definition of CAT $\lceil\kappa\rceil$ spaces (see [1]). Namely, a metric space $\mathcal{U}$ in which any two points at distance $< \varpi^\kappa$ are joined by a geodesic is a CAT $\lceil\kappa\rceil$ space if and only if it satisfies the following condition:

(1) (point-on-side comparison) for any geodesic $[xy]$ and $z \in]xy[$, we have

$$\tilde\measuredangle^\kappa(x\,{}^p_y) \geqslant \tilde\measuredangle^\kappa(x\,{}^p_z),$$

or equivalently,

$$|\tilde p\tilde z| \geqslant |pz|, \qquad ❹$$

where $[\tilde p\tilde x\tilde y] = \tilde\triangle^\kappa(pxy)$, $\tilde z \in]\tilde x\tilde y[$, $|\tilde x\tilde z| = |xz|$.

We also use Reshetnyak's majorization theorem [12]. Suppose $\tilde\alpha$ is a simple closed curve of finite length in $\mathbb{M}^2[\kappa]$, and $D \subset \mathbb{M}^2[\kappa]$ is a closed region bounded by $\tilde\alpha$. If $\mathcal{X}$ is a metric space, a length-nonincreasing map $F\colon D \to \mathcal{X}$ is called *majorizing* if it is length-preserving on $\tilde\alpha$. In this case, we say that D *majorizes* the curve $\alpha = F \circ \tilde\alpha$ under the map F.

2.5. Reshetnyak's majorization theorem. *Any closed curve α of length $< 2\cdot\varpi^\kappa$ in $\mathcal{U} \in$ CAT $\lceil\kappa\rceil$ is majorized by a convex region in $\mathbb{M}^2[\kappa]$.*

Ultralimit of metric spaces. Given a metric space $\mathcal{X}$, its ultrapower (i.e., ultralimit of constant sequence $\mathcal{X}_n = \mathcal{X}$) will be denoted as $\mathcal{X}^\omega$; here ω denotes a fixed nonprinciple ultrafilter. For definitions and properties of ultrapowers, we refer to a paper of Kleiner and Leeb [8, 2.4].

We use the following facts about ultrapowers which easily follow from the definitions (see [1] for details):

- ◇ $\mathcal{X} \in \overline{\text{CAT}}\lceil\kappa\rceil \iff \mathcal{X}^\omega \in \overline{\text{CAT}}\lceil\kappa\rceil$.
- ◇ $\mathcal{X} \in \overline{\text{CBB}}\lfloor\kappa\rfloor \iff \mathcal{X}^\omega \in \overline{\text{CBB}}\lfloor\kappa\rfloor$.
- ◇ $\mathcal{X}$ is intrinsic if and only if $\mathcal{X}^\omega$ is geodesic.

Note that if $\mathcal{X}$ is *proper* (namely, bounded closed sets are compact), then $\mathcal{X}$ and $\mathcal{X}^\omega$ coincide. Thus a reader interested only in proper spaces may ignore everything related to ultrapower in this article.

3. Short map extension definitions.

Theorems 3.1 and 3.2 give characterizations of $\overline{\text{CBB}}\lfloor\kappa\rfloor$ and CAT $\lceil\kappa\rceil$. Very similar theorems were proved by Lang and Shroeder in [9].

3.1. Theorem. *Let $\mathcal{L}$ be a complete intrinsic space. Then $\mathcal{L} \in \overline{\text{CBB}}\lfloor\kappa\rfloor$ if and only if for any 3-point set V_3 and any 4-point set $V_4 \supset V_3$ in $\mathcal{L}$, any short map $f\colon V_3 \to \mathbb{M}^2[\kappa]$ can be extended to a short map $F\colon V_4 \to \mathbb{M}^2[\kappa]$ (so $f = F|_{V_3}$).*

3.2. Theorem. *Let $\mathcal{U}$ be a metric space in which any pair of points at distance $< \varpi^\kappa$ are joined by a unique geodesic. Then $\mathcal{U} \in$ CAT $\lceil\kappa\rceil$ if and only if for any 3-point set V_3 and 4-point set $V_4 \supset V_3$ in $\mathbb{M}^2[\kappa]$, where the perimeter of V_3 is $< 2\cdot\varpi^\kappa$, any short map $f\colon V_3 \to \mathcal{U}$ can be extended to a short map $F\colon V_4 \to \mathcal{U}$.*

The proof of the "only if" part of Theorem 3.1 can be obtained as a corollary of Kirszbraun's theorem (5.1). But we present another proof, based on more elementary ideas. The "only if" part of Theorem 3.2 does not follow directly from Kirszbraun's theorem, since the desired extension is in $\mathcal{U}$, not just the completion of $\mathcal{U}$.

In the proof of Theorem 3.1, we use the following lemma in the geometry of model planes. Here we say that two triangles with a common vertex *do not overlap* if their convex hulls intersect only at the common vertex.

3.3. Overlap lemma. *Let $[\tilde x^1\tilde x^2\tilde x^3]$ be a triangle in $\mathbb{M}^2[\kappa]$. Let $\tilde p^1,\tilde p^2,\tilde p^3$ be points such that, for any permutation $\{i,j,k\}$ of $\{1,2,3\}$, we have*

(i) $|\tilde p^i\tilde x^k| = |\tilde p^j\tilde x^k|$,

(ii) $\tilde p^i$ and $\tilde x^i$ lie in the same closed halfspace determined by $[\tilde x^j\tilde x^k]$,

(iii) $\measuredangle[\tilde x^i\,{}^{\tilde x^j}_{\tilde p^k}] + \measuredangle[\tilde x^i\,{}^{\tilde p^j}_{\tilde x^k}] < \pi$.

Set $\measuredangle\tilde p^i = \measuredangle[\tilde p^i\,{}^{\tilde x^k}_{\tilde x^j}]$. It follows that:

a) *If $\measuredangle\tilde p^1 + \measuredangle\tilde p^2 + \measuredangle\tilde p^3 \leqslant 2\cdot\pi$ and triangles $[\tilde p^3\tilde x^1\tilde x^2]$, $[\tilde p^2\tilde x^3\tilde x^1]$ do not overlap, then*

$$\measuredangle\tilde p^1 > \measuredangle\tilde p^2 + \measuredangle\tilde p^3.$$

b) *No pair of triangles $[\tilde p^i\tilde x^j\tilde x^k]$ overlap if and only if*

$$\measuredangle\tilde p^1 + \measuredangle\tilde p^2 + \measuredangle\tilde p^3 > 2\cdot\pi.$$

Remark. If $\kappa \leqslant 0$, the "only if" part of (b) can be proved without using condition (i). This follows immediately from the formula that relates the sum of angles for the hexagon $[\tilde p^1\tilde x^2\tilde p^3\tilde x^1\tilde p^2\tilde x^3]$ and its area:

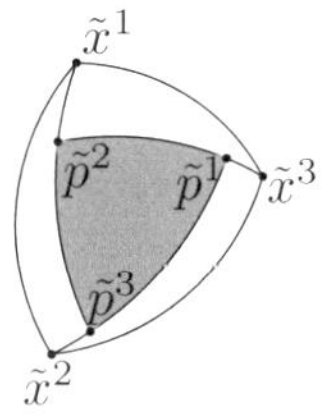

$$\measuredangle\tilde p^1 - \measuredangle\tilde x^2 + \measuredangle p^3 - \measuredangle\tilde x^1 + \measuredangle\tilde p^2 - \measuredangle\tilde x^3 = 2\cdot\pi - \kappa\cdot\text{area}.$$

In the case $\kappa > 0$, condition (i) is essential. An example for $\kappa > 0$ can be constructed by perturbing the degenerate spherical configuration on the picture.

Proof. Rotate the triangle $[\tilde p^3\tilde x^1\tilde x^2]$ around $\tilde x^1$ to make $[\tilde x^1\tilde p^3]$ coincide with $[\tilde x^1\tilde p^2]$. Let $\dot x^2$ denote the image of $\tilde x^2$ after rotation. By (ii) and (iii), the triangles $[\tilde p^3\tilde x^1\tilde x^2]$ and $[\tilde p^2\tilde x^3\tilde x^1]$ do not overlap if and only if $\measuredangle[\tilde x^1\,{}^{\tilde x^3}_{\dot x^2}] < \measuredangle[\tilde x^1\,{}^{\tilde x^3}_{\tilde x^2}]$, and hence if and only if $|\dot x^2\tilde x^3| < |\tilde x^2\tilde x^3|$. This inequality holds if and only if

$$\begin{aligned}\measuredangle\tilde p^1 &> \measuredangle[\tilde p^2\,{}^{\tilde x^3}_{\dot x^2}]\\ &= \min\{\measuredangle\tilde p^3 + \measuredangle\tilde p^2, 2\cdot\pi - (\measuredangle\tilde p^3 + \measuredangle\tilde p^2)\},\end{aligned} \qquad ❶$$

since in the inequality, the corresponding hinges have the same pairs of sidelengths. (The two pictures show that both possibilities for the minimum can occur.)

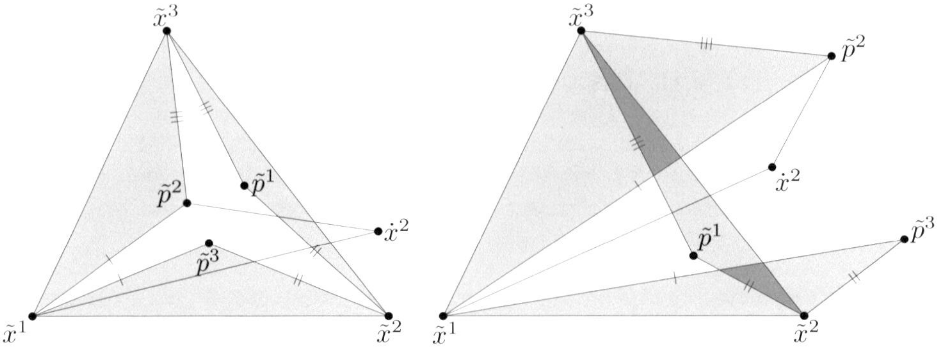

If $\measuredangle\tilde p^1+\measuredangle\tilde p^2+\measuredangle\tilde p^3\leqslant 2\cdot\pi$, then ❶ implies $\measuredangle\tilde p^1>\measuredangle\tilde p^2+\measuredangle\tilde p^3$. That proves (a).

"Only if" part of (b). Suppose no two triangles overlap and $\measuredangle\tilde p^2+\measuredangle\tilde p^2+\measuredangle\tilde p^3\leqslant 2\cdot\pi$. By (a), for $\{i,j,k\}=\{1,2,3\}$ we have

$$\measuredangle\tilde p^i>\measuredangle\tilde p^j+\measuredangle\tilde p^k.$$

Adding these three inequalities gives a contradiction:

$$\measuredangle\tilde p^1+\measuredangle\tilde p^2+\measuredangle\tilde p^3>2\cdot(\measuredangle\tilde p^1+\measuredangle\tilde p^2+\measuredangle\tilde p^3).$$

"If" part of (b). Suppose triangles $[\tilde p^3\tilde x^1\tilde x^2]$ and $[\tilde p^2\tilde x^3\tilde x^1]$ overlap and

$$\measuredangle\tilde p^1+\measuredangle\tilde p^2+\measuredangle\tilde p^3>2\cdot\pi. \qquad ❷$$

By the former, ❶ fails. By ❷, $\measuredangle\tilde p^2+\measuredangle\tilde p^3>\pi$. Therefore

$$\measuredangle\tilde p^1\leqslant 2\cdot\pi-(\measuredangle\tilde p^2+\measuredangle\tilde p^3),$$

which contradicts ❷. □

Proof of 3.1; "if" part. Assume $\mathcal{L}$ is geodesic. Let $x^1,x^2,x^3\in\mathcal{L}$ be such that the model triangle $[\tilde x^1\tilde x^2\tilde x^3]=\tilde\triangle^\kappa(x^1x^2x^3)$ is defined. Choose $p\in\,]x^1x^2[$. Let $V_3=\{x^1,x^2,x^3\}$ and $V_4=\{x^1,x^2,x^3,p\}$, and set $f(x^i)=\tilde x^i$. Then a short extension of f to V_4 gives point-on-side comparison (see page 91).

In the case when $\mathcal{L}$ is not geodesic, pass to its ultrapower $\mathcal{L}^\omega$. Note that if $\mathcal{L}$ satisfies the conditions of Theorem 3.1 then so does $\mathcal{L}^\omega$. Also, recall that $\mathcal{L}^\omega$ is geodesic. Thus, from above, $\mathcal{L}^\omega\in\overline{\mathrm{CBB}}[\kappa]$. Hence $\mathcal{L}\in\overline{\mathrm{CBB}}[\kappa]$.

"Only if" part. Assume the contrary; i.e., $x^1,x^2,x^3,p\in\mathcal{L}$, and $\tilde x^1,\tilde x^2,\tilde x^3\in\mathbb{M}^2[\kappa]$ are such that $|\tilde x^i\tilde x^j|\leqslant|x^ix^j|$ for all i,j and there is no point $\tilde p\in\mathbb{M}^2[\kappa]$ such that $|\tilde p\tilde x^i|\leqslant|px^i|$ for all i.

We claim that in this case all comparison triangles $\tilde{\triangle}^\kappa(px^ix^j)$ are defined. That is always true if $\kappa \leqslant 0$. If $\kappa > 0$, and say $\tilde{\triangle}^\kappa(px^1x^2)$ is undefined, then

$$\begin{aligned}|px^1| + |px^2| &\geqslant 2\cdot\varpi^\kappa - |x^1x^2| \geqslant \\ &\geqslant 2\cdot\varpi^\kappa - |\tilde{x}^1\tilde{x}^2| \geqslant \\ &\geqslant |\tilde{x}^1\tilde{x}^3| + |\tilde{x}^2\tilde{x}^3|.\end{aligned}$$

Thus one can take $\tilde{p}$ on $[\tilde{x}^1\tilde{x}^3]$ or $[\tilde{x}^2\tilde{x}^3]$.

For each $i \in \{1,2,3\}$, consider a point $\tilde{p}^i \in \mathbb{M}^2[\kappa]$ such that $|\tilde{p}^i\tilde{x}^i|$ is minimal among points satisfying $|\tilde{p}^i\tilde{x}^j| \leqslant |px^j|$ for all $j \neq i$. Clearly, every $\tilde{p}^i$ is inside the triangle $[\tilde{x}^1\tilde{x}^2\tilde{x}^3]$ (that is, in $\mathrm{Conv}(\tilde{x}^1,\tilde{x}^2,\tilde{x}^3)$), and $|\tilde{p}^i\tilde{x}^i| > |px^i|$ for each i. It follows that

(i) $|\tilde{p}^i\tilde{x}^j| = |px^j|$ for $i \neq j$;

(ii) no pair of triangles from $[\tilde{p}^1\tilde{x}^2\tilde{x}^3]$, $[\tilde{p}^2\tilde{x}^3\tilde{x}^1]$, $[\tilde{p}^3\tilde{x}^1\tilde{x}^2]$ overlap in $[\tilde{x}^1\tilde{x}^2\tilde{x}^3]$.

As follows from Lemma 3.3*b*, in this case

$$\measuredangle[\tilde{p}^1\,{}^{\tilde{x}^2}_{\tilde{x}^3}] + \measuredangle[\tilde{p}^2\,{}^{\tilde{x}^3}_{\tilde{x}^1}] + \measuredangle[\tilde{p}^3\,{}^{\tilde{x}^1}_{\tilde{x}^2}] > 2\cdot\pi.$$

Thus we arrive at a contradiction, since $|\tilde{x}^i\tilde{x}^j| \leqslant |x^ix^j|$ implies that

$$\measuredangle[\tilde{p}^k\,{}^{\tilde{x}^i}_{\tilde{x}^j}] \leqslant \tilde{\measuredangle}^\kappa(p\,{}^{x^i}_{x^j})$$

if (i,j,k) is a permutation of $(1,2,3)$. □

In the proof of Theorem 3.2, we use the following lemma in the geometry of model planes:

3.4. Lemma. *Let $x^1,x^2,x^3,y^1,y^2,y^3 \in \mathbb{M}[\kappa]$ be points such that $|x^ix^j| \geqslant |y^iy^j|$ for all i,j. Then there is a short map $\Phi\colon \mathbb{M}[\kappa] \to \mathbb{M}[\kappa]$ such that $\Phi(x^i) = y^i$ for all i; moreover, one can choose Φ so that*

$$\mathrm{Im}\,\Phi \subset \mathrm{Conv}(y^1,y^2,y^3).$$

We only give an idea of the proof of this lemma; alternatively, one can get the result as a corollary of Kirszbraun's theorem (5.1)

Idea of the proof. The map Φ can be constructed as a composition of the following folding maps: Given a halfspace H in $\mathbb{M}[\kappa]$, consider the map $\mathbb{M}[\kappa] \to H$, which is the identity on H and reflects all points outside of H into H. This map is a path isometry, in particular, it is short.

One can get the last part of the lemma by composing the above map with foldings along the sides of the triangle $[y^1y^2y^3]$ and passing to a partial limit. □

Proof of 3.2; "if" part. The point-on-side comparison (1) follows by taking $V_3 = \{\tilde{x},\tilde{y},\tilde{p}\}$ and $V_4 = \{\tilde{x},\tilde{y},\tilde{p},\tilde{z}\}$ where $z \in\,]xy[$. It is only necessary to observe that $F(\tilde{z}) = z$ by uniqueness of $[xy]$.

"Only if" part. Let $V_3 = \{\tilde{x}^1,\tilde{x}^2,\tilde{x}^3\}$ and $V_4 = \{\tilde{x}^1,\tilde{x}^2,\tilde{x}^3,\tilde{p}\}$.

Set $y^i = f(\tilde{x}^i)$ for all i; we need to find a point $q \in \mathcal{U}$ such that $|y^iq| \leqslant |\tilde{x}^i\tilde{p}|$ for all i.

Consider the model triangle $[\tilde{y}^1\tilde{y}^2\tilde{y}^3] = \tilde{\triangle}^\kappa(y^1y^2y^3)$. Set $D = \mathrm{Conv}(\tilde{y}^1,\tilde{y}^2,\tilde{y}^3)$.

Note that $|\tilde{y}^i\tilde{y}^j| = |y^iy^j| \leqslant |\tilde{x}^i\tilde{x}^j|$ for all i,j. Applying Lemma 3.4, we get a short map $\Phi\colon \mathbb{M}[\kappa] \to D$ such that $\Phi\colon \tilde{x}^i \mapsto \tilde{y}^i$.

Further, from Reshetnyak majorization (2.5), there is a short map $F\colon D \to \mathcal{U}$ such that $\tilde{y}^i \mapsto y^i$ for all i.

Thus one can take $q = F \circ \Phi(\tilde{p})$. □

4. (1+n)-point comparison

The following theorem gives a more sensitive analog of (1+3)-point comparison. In a bit more analytic form it was discovered by Sturm in [13].

4.1. (1+n)-point comparison. *Let $\mathcal{L} \in \overline{\mathrm{CBB}}[\kappa]$. Then for any array of points $p, x^1, \dots, x^n \in \mathcal{L}$ there is a model array $\tilde{p}, \tilde{x}^1, \dots, \tilde{x}^n \in \mathbb{M}^n[\kappa]$ such that*

a) *$|\tilde{p}\tilde{x}^i| = |px^i|$ for all i.*

b) *$|\tilde{x}^i\tilde{x}^j| \geqslant |x^ix^j|$ for all i,j.*

Proof. It is enough to show that given $\varepsilon > 0$ there is a configuration $\tilde{p}, \tilde{x}^1, \dots, \tilde{x}^n$ in $\mathbb{M}^n[\kappa]$ such that $|\tilde{x}^i\tilde{x}^j| \geqslant |x^ix^j|$ and $\big||\tilde{p}\tilde{x}^i| - |px^i|\big| \leqslant \varepsilon$. Then one can pass to a limit configuration for $\varepsilon \to 0+$.

According to 2.1, there is a point p' such that $|p'p| \leqslant \varepsilon$ and $\mathrm{T}_{p'}$ contains a subcone E isometric to a Euclidean space which contains all vectors $\log[p'x^i]$. Passing to a subspace if necessary, we can assume that $\dim E \leqslant n$.

Mark a point $\tilde{p} \in \mathbb{M}^n[\kappa]$ and choose an isometric embedding $\imath\colon E \to \mathrm{T}_{\tilde{p}}\mathbb{M}^n[\kappa]$. Set

$$\tilde{x}^i = \exp_{\tilde{p}} \circ \imath \circ \log[p'x^i].$$

Thus $|\tilde{p}\tilde{x}^i| = |p'x^i|$ and therefore

$$\big||\tilde{p}\tilde{x}^i| - |px^i|\big| \leqslant |pp'| \leqslant \varepsilon.$$

From the hinge comparison, we have

$$\tilde{\measuredangle}^\kappa(\tilde{p}\,{}^{\tilde{x}^i}_{\tilde{x}^j}) = \measuredangle[\tilde{p}\,{}^{\tilde{x}^i}_{\tilde{x}^j}] = \measuredangle[p'\,{}^{x^i}_{x^j}] \geqslant \tilde{\measuredangle}^\kappa(p'\,{}^{x^i}_{x^j}),$$

thus

$$|\tilde{x}^i\tilde{x}^j| \geqslant |x^ix^j|.$$ □

5. Kirszbraun's theorem

A slightly weaker version of the following theorem was proved by Lang and Schroeder in [9]. The Conjecture 7.4 (if true) gives an equivalent condition for the existence of a short extension; roughly it states that example 5.2 is the only obstacle.

5.1. Kirszbraun's theorem. *Let $\mathcal{L} \in \overline{\mathrm{CBB}}[\kappa]$, $\mathcal{U} \in \overline{\mathrm{CAT}}[\kappa]$, $Q \subset \mathcal{L}$ be an arbitrary subset and $f\colon Q \to \mathcal{U}$ be a short map. Assume that there is $z \in \mathcal{U}$ such that $f(Q)$ is in $\mathrm{B}[z, \frac{\varpi^\kappa}{2}]$. Then $f\colon Q \to \mathcal{U}$ can be extended to a short map $F\colon \mathcal{L} \to \mathcal{U}$ (that is, there is a short map $F\colon \mathcal{L} \to \mathcal{U}$ such that $F|_Q = f$.)*

The condition $f(Q) \subset \mathrm{B}[z, \frac{\varpi^\kappa}{2}]$ trivially holds for any $\kappa \leqslant 0$ since in this case $\varpi^\kappa = \infty$. The following example shows that this condition is needed for $\kappa > 0$.

5.2. Example. *Let $\mathbb{S}^m_+$ be a closed m-dimensional unit hemisphere. Denote its boundary, which is isometric to $\mathbb{S}^{m-1}$, by $\partial\mathbb{S}^m_+$. Clearly, $\mathbb{S}^m_+ \in \overline{\mathrm{CBB}}\lfloor 1\rfloor$ and $\partial\mathbb{S}^m_+ \in \overline{\mathrm{CAT}}\lceil 1\rceil$ but the identity map $\partial\mathbb{S}^m_+ \to \partial\mathbb{S}^m_+$ cannot be extended to a short map $\mathbb{S}^m_+ \to \partial\mathbb{S}^m_+$ (there is no place for the pole).*

There is also a direct generalization of this example to a hemisphere in a Hilbert space of arbitrary cardinal dimension.

First we prove this theorem in the case $\kappa \leqslant 0$ (5.4). In the proof of the more complicated case $\kappa > 0$, we use the case $\kappa = 0$. The following lemma is the main ingredient in the proof.

5.3. Finite+one lemma. *Let $\kappa \leqslant 0$, $\mathcal{L} \in \overline{\mathrm{CBB}}\lfloor\kappa\rfloor$, and $\mathcal{U} \in \overline{\mathrm{CAT}}\lceil\kappa\rceil$. Moreover, let $x^1, x^2, \dots, x^n \in \mathcal{L}$ and $y^1, y^2, \dots, y^n \in \mathcal{U}$ be such that $|x^i x^j| \geqslant |y^i y^j|$ for all i, j.*

Then for any $p \in \mathcal{L}$, there is a $q \in \mathcal{U}$ such that $|y^i q| \leqslant |x^i p|$ for each i.

Proof. It is sufficient to prove the lemma only for $\kappa = 0$ and -1. The proofs of these two cases are identical, only the formulas differ. In the proof, we assume $\kappa = 0$ and provide the formulas for $\kappa = -1$ in the footnotes.

From (1+n)-point comparison (4.1), there is a model configuration $\tilde{p}, \tilde{x}^1, \tilde{x}^2, \dots, \tilde{x}^n$ in $\mathbb{M}^n[\kappa]$ such that $|\tilde{p}\tilde{x}^i| = |px^i|$ and $|\tilde{x}^i\tilde{x}^j| \geqslant |x^i x^j|$ for all i, j.

For each i, consider functions $f^i\colon \mathcal{U} \to \mathbb{R}$ and $\tilde{f}^i\colon \mathbb{M}^n[\kappa] \to \mathbb{R}$ defined as follows[1]:

$$f^i(y) = \tfrac{1}{2}\cdot|y^i y|^2, \quad \tilde{f}^i(\tilde{x}) = \tfrac{1}{2}\cdot|\tilde{x}^i\tilde{x}|^2. \qquad (A)^0$$

Set $\boldsymbol{f} = (f^1, f^2, \dots, f^n)\colon \mathcal{U} \to \mathbb{R}^n$ and $\tilde{\boldsymbol{f}} = (\tilde{f}^1, \tilde{f}^2, \dots, \tilde{f}^n)\colon \mathbb{M}^n[\kappa] \to \mathbb{R}^n$.

Recall that SupSet (superset in $\mathbb{R}^n$) is defined in A.1. Note that it is sufficient to prove that $\tilde{\boldsymbol{f}}(\tilde{p}) \in \operatorname{SupSet} \boldsymbol{f}(\mathcal{U})$.

Clearly, $(f^i)'' \geqslant 1$. Thus, by the theorem on barycentric simplex (A.2*b*), the set $\operatorname{SupSet} \boldsymbol{f}(\mathcal{U}) \subset \mathbb{R}^n$ is convex.

Arguing by contradiction, let us assume that $\tilde{\boldsymbol{f}}(\tilde{p}) \notin \operatorname{SupSet} \boldsymbol{f}(\mathcal{U})$.

Then there exists a supporting hyperplane $\alpha_1 x_1 + \dots \alpha_n x_n = c$ to $\operatorname{SupSet} \boldsymbol{f}(\mathcal{U})$, separating it from $\tilde{\boldsymbol{f}}(\tilde{p})$. Just as in the proof of Theorem A.2 we have that all $\alpha_i \geqslant 0$. So by rescaling we can assume that $(\alpha_1, \alpha_2, \dots, \alpha_n) \in \Delta^{n-1}$ and

$$\sum_i \alpha_i \cdot \tilde{f}^i(\tilde{p}) < \inf\left\{ \sum_i \alpha_i \cdot f^i(q) \,\middle|\, q \in \mathcal{U} \right\}.$$

The latter contradicts the following claim.

[1] In case $\kappa = -1$,

$$f^i(y) = \cosh|y^i y|, \quad \tilde{f}^i(\tilde{x}) = \cosh|\tilde{x}^i\tilde{x}|. \qquad (A)^-$$

I. Claim. *Given* $\boldsymbol{\alpha} = (\alpha_1, \alpha_2, \dots, \alpha_n) \in \Delta^{n-1}$, *set*

$$h = \sum_i \alpha_i \cdot f^i, \qquad h\colon \mathcal{U} \to \mathbb{R}, \qquad z = \operatorname{argmin} h \in \mathcal{U},$$

$$\tilde h = \sum_i \alpha_i \cdot \tilde f^i, \qquad \tilde h\colon \mathbb{M}^n[\kappa] \to \mathbb{R}, \qquad \tilde z = \operatorname{argmin} \tilde h \in \mathbb{M}^n[\kappa].$$

Then $h(z) \leqslant \tilde h(\tilde z)$.

Proof of the claim. Note that $\boldsymbol{d}_z h \geqslant 0$. Thus, for each i, we have[2]

$$\begin{aligned} 0 &\leqslant (\boldsymbol{d}_z h)(\operatorname{dir}[z y^i]) = \\ &= -\sum_j \alpha_j \cdot |z y^j| \cdot \cos \measuredangle[z\,{}^{y^i}_{y^j}] \leqslant \\ &\leqslant -\sum_j \alpha_j \cdot |z y^j| \cdot \cos \tilde\measuredangle^0(z\,{}^{y^i}_{y^j}) = \\ &= -\tfrac{1}{2\cdot|zy^i|} \cdot \sum_j \alpha_j \cdot \left[|zy^i|^2 + |zy^j|^2 - |y^i y^j|^2\right]. \end{aligned} \qquad (B)^0$$

In particular[3],

$$\sum_i \alpha_i \cdot \left[\sum_j \alpha_j \cdot \left[|zy^i|^2 + |zy^j|^2 - |y^i y^j|^2\right]\right] \leqslant 0, \qquad (C)^0$$

or[4]

$$2\cdot h(z) \leqslant \sum_{i,j} \alpha_i \cdot \alpha_j \cdot |y^i y^j|^2 \qquad (D)^0$$

---

[2]In case $\kappa = -1$, the same calculations give

$$0 \leqslant \dots \leqslant -\tfrac{1}{\sinh|zy^i|} \cdot \sum_j \alpha_j \cdot \left[\cosh|zy^i| \cdot \cosh|zy^j| - \cosh|y^i y^j|\right]. \qquad (B)^-$$

[3]In case $\kappa = -1$, the same calculations give

$$\sum_i \alpha_i \cdot \left[\sum_j \alpha_j \cdot \left[\cosh|zy^i| \cdot \cosh|zy^j| - \cosh|y^i y^j|\right]\right] \leqslant 0. \qquad (C)^-$$

[4]In case $\kappa = -1$,

$$(h(z))^2 \leqslant \sum_{i,j} \alpha_i \cdot \alpha_j \cdot \cosh|y^i y^j|. \qquad (D)^-$$

Note that if $\mathcal{U} \overset{iso}{=\!=} \mathbb{M}^n[\kappa]$, then all inequalities in (B, C, D) are sharp. Thus the same argument as above, repeated for $\tilde{x}^1, \tilde{x}^2, \ldots, \tilde{x}^n \in \mathbb{M}^n[\kappa]$ gives[5]

$$2{\cdot}\tilde{h}(\tilde{z}) = \sum_{i,j} \alpha_i{\cdot}\alpha_j{\cdot}|\tilde{x}^i\tilde{x}^j|^2. \qquad (E)^0$$

Note that

$$|\tilde{x}^i\tilde{x}^j| \geqslant |x^i x^j| \geqslant |y^i y^j|$$

for all i, j. Thus, (D) and (E) imply the claim. □

5.4. Kirszbraun's theorem for nonpositive bound. *Let $\kappa \leqslant 0$, $\mathcal{L} \in \overline{\mathrm{CBB}}\lfloor\kappa\rfloor$, $\mathcal{U}$ is in $\overline{\mathrm{CAT}}\lceil\kappa\rceil$, Q is an arbitrary subset of $\mathcal{L}$ and $f\colon Q \to \mathcal{U}$ be a short map. Then there is a short extension $F\colon \mathcal{L} \to \mathcal{U}$ of f; that is, there is a short map $F\colon \mathcal{L} \to \mathcal{U}$ such that $F|_Q = f$.*

Remark. If $\mathcal{U}$ is proper, then in the following proof Helly's theorem (B.1) is not needed. Everything follows directly from compactness of closed balls in $\mathcal{U}$.

Proof of 5.4. By Zorn's lemma, we can assume that $Q \subset \mathcal{L}$ is a maximal set; i.e., $f\colon Q \to \to \mathcal{U}$ does not admit a short extension to any larger set $Q' \supset Q$.

Let us argue by contradiction. Assume that $Q \neq \mathcal{L}$; choose $p \in \mathcal{L}\backslash Q$. Then

$$\bigcap_{x\in Q} \overline{\mathrm{B}}[f(x), |px|] = \varnothing.$$

Since $\kappa \leqslant 0$, the balls are convex; thus, by Helly's theorem (B.1), one can choose a point array $x^1, x^2, \ldots, x^n \in Q$ such that

$$\bigcap_{i=1}^{n} \overline{\mathrm{B}}[y^i, |x^i p|] = \varnothing, \qquad ❶$$

where $y^i = f(x^i)$. Finally note that ❶ contradicts the Finite+one lemma (5.3). □

Proof of Kirszbraun's theorem (5.1). The case $\kappa \leqslant 0$ is already proved in 5.4. Thus it remains to prove the theorem only in case $\kappa > 0$. After rescaling we may assume that $\kappa = 1$ and therefore $\varpi^\kappa = \pi$.

Since $\overline{\mathrm{B}}[z, \pi/2] \in \overline{\mathrm{CAT}}\lceil\kappa\rceil$ (2.4, 2.2), we can assume $\mathcal{U} = \overline{\mathrm{B}}[z, \pi/2]$. In particular, any two points of $\mathcal{U}$ at distance $< \pi$ are joined by a geodesic, and $\operatorname{diam}\mathcal{U} \leqslant \pi$. If $|xy| = \pi$ for some $x, y \in \mathcal{U}$, then the concatenation of $[xz]$ and $[zy]$ forms a geodesic $[xy]$. Hence $\mathcal{U}$ is geodesic.

Further, we can also assume that $\operatorname{diam}\mathcal{L} \leqslant \pi$. Otherwise $\mathcal{L}$ is one-dimensional; in this case the result follows since $\mathcal{U}$ is geodesic.

---

[5] In the case $\kappa = -1$,

$$(\tilde{h}(\tilde{z}))^2 = \sum_{i,j} \alpha_i{\cdot}\alpha_j{\cdot}\cosh|\tilde{x}^i\tilde{x}^j|. \qquad (E)^-$$

Assume the theorem is false. Then there is a set $Q\subset\mathcal{L}$, a short map $f\colon Q\to\mathcal{U}$ and $p\in\mathcal{L}\backslash Q$ such that

$$\bigcap_{x\in Q}\overline{\mathrm{B}}[f(x),|xp|]=\varnothing. \qquad ❷$$

We are going to apply 5.4 for $\kappa=0$ to the Euclidean cones $\mathring{\mathcal{L}}=\mathrm{Cone}\,\mathcal{L}$ and $\mathring{\mathcal{U}}=\mathrm{Cone}\,\mathcal{U}$. Note that

◇ $\mathring{\mathcal{U}}\in\overline{\mathrm{CAT}}[0]$,

◇ since $\operatorname{diam}\mathcal{L}\leqslant\pi$ we have $\mathring{\mathcal{L}}\in\overline{\mathrm{CBB}}[0]$.

Further, we view the spaces $\mathcal{L}$ and $\mathcal{U}$ as unit spheres in $\mathring{\mathcal{L}}$ and $\mathring{\mathcal{U}}$ respectively. In the cones $\mathring{\mathcal{L}}$ and $\mathring{\mathcal{U}}$, we use "$|{*}|$" for distance to the vertex, say o, and "$\cdot$" for cone multiplication. We also use short-cuts $\measuredangle(x,y)\stackrel{def}{=\!=}\measuredangle[o\,{}^{x}_{y}]$ and

$$\begin{aligned}\langle x,y\rangle &\stackrel{def}{=\!=}|x|\cdot|y|\cdot\cos\measuredangle[o\,{}^{x}_{y}]=\\ &=\tfrac12\left(|x|^2+|y|^2-|xy|^2\right).\end{aligned}$$

In particular,

◇ $|xy|_{\mathcal{L}}=\measuredangle(x,y)$ for any $x,y\in\mathcal{L}$,

◇ $|xy|_{\mathcal{U}}=\measuredangle(x,y)$ for any $x,y\in\mathcal{U}$,

◇ for any $y\in\mathcal{U}$, we have

$$\measuredangle(z,y)\leqslant\tfrac{\pi}{2}. \qquad ❸$$

Set $\mathring{Q}=\mathrm{Cone}\,Q\subset\mathring{\mathcal{L}}$ and let $\mathring{f}\colon\mathring{Q}\to\mathring{\mathcal{U}}$ be the natural cone extension of f; i.e., $y=f(x)$ $\Rightarrow t\cdot y=\mathring{f}(t\cdot x)$ for $t\geqslant 0$. Clearly $\mathring{f}$ is short.

Applying 5.4 for $\mathring{f}$, we get a short extension map $\mathring{F}\colon\mathring{\mathcal{L}}\to\mathring{\mathcal{U}}$. Set $s=\mathring{F}(p)$. Thus,

$$|s\mathring{f}(w)|\leqslant|pw| \qquad ❹$$

for any $w\in\mathring{Q}$. In particular, $|s|\leqslant 1$. Applying ❹ for $w=t\cdot x$ and $t\to\infty$ we get

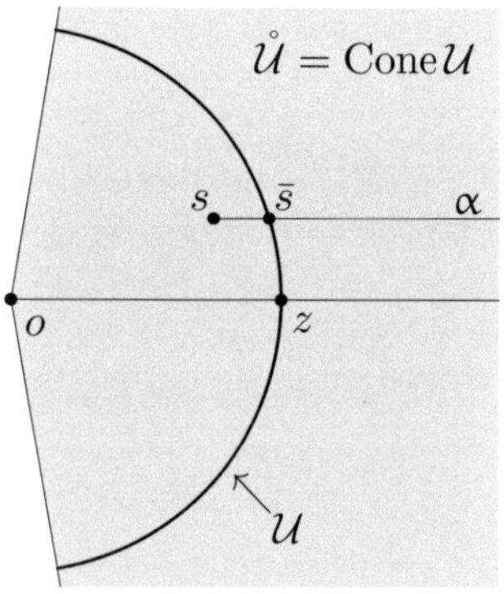

$$\langle f(x),s\rangle\geqslant\cos\measuredangle(p,x) \qquad ❺$$

for any $x\in Q$.

Since $\mathcal{U}\in\overline{\mathrm{CAT}}[0]$, the geodesics $\mathrm{geod}_{[s\ t\cdot z]}$ converge as $t\to\infty$ to a ray, say $\alpha\colon[0,\infty)\to\mathring{\mathcal{U}}$. From ❸, we have that the function $t\mapsto\langle f(x),\alpha(t)\rangle$ is non-decreasing. Therefore, from ❺, for the necessarily unique point $\bar{s}$ on the ray α such that $|\bar{s}|=1$ we also have

$$\langle f(x),\bar{s}\rangle\geqslant\cos\measuredangle(p,x)$$

or

$$\measuredangle(\bar{s},f(x))\leqslant\measuredangle(p,f(x))$$

for any $x\in Q$. The latter contradicts ❷. □

6. (2n+2)-point comparison

Here we give a generalization of the (2+2)-point comparison to (2n+2) points. It follows from the generalized Kirszbraun's theorem.

First let us give a reformulation of (2+2)-point comparison.

6.1. Reformulation of (2+2)-point comparison. *Let $\mathcal{X}$ be a metric space. A quadruple $p, q, x, y \in \mathcal{X}$ satisfies (2+2)-point comparison if one of the following holds:*

a) *One of the triples (p, q, x) or (p, q, y) has perimeter $> 2\cdot\varpi^\kappa$.*

b) *If $[\tilde{p}\tilde{q}\tilde{x}] = \tilde{\triangle}^\kappa(pqx)$ and $[\tilde{p}\tilde{q}\tilde{y}] = \tilde{\triangle}^\kappa pqy$, then*

$$|\tilde{x}\tilde{z}| + |\tilde{z}\tilde{y}| \geqslant |xy|,$$

for any $\tilde{z} \in [\tilde{p}\tilde{q}]$.

6.2. (2n+2)-point comparison. *Let $\mathcal{U} \in \overline{\mathrm{CAT}}[\kappa]$. Consider $x, y \in \mathcal{U}$ and an array of pairs of points $(p^1, q^1), (p^2, q^2), \ldots, (p^n, q^n)$ in $\mathcal{U}$, such that there is a model configuration $\tilde{x}, \tilde{y}$ and array of pairs $(\tilde{p}^1, \tilde{q}^1), (\tilde{p}^2, \tilde{q}^2), \ldots, (\tilde{p}^n, \tilde{q}^n)$ in $\mathbb{M}^3[\kappa]$ with the following properties:*

a) *$[\tilde{x}\tilde{p}^1\tilde{q}^1] = \tilde{\triangle}^\kappa xp^1q^1$ and $[\tilde{y}\tilde{p}^n\tilde{q}^n] = \tilde{\triangle}^\kappa yp^nq^n$;*

b) *The simplex $\tilde{p}^i\tilde{p}^{i+1}\tilde{q}^i\tilde{q}^{i+1}$ is a model simplex[6] of $p^ip^{i+1}q^iq^{i+1}$ for all i.*

Then for any choice of n points $\tilde{z}^i \in [\tilde{p}^i\tilde{q}^i]$, we have

$$|\tilde{x}\tilde{z}^1| + |\tilde{z}^1\tilde{z}^2| + \ldots + |\tilde{z}^{n-1}\ \tilde{z}^n| + |\tilde{z}^n\tilde{y}| \geqslant |xy|.$$

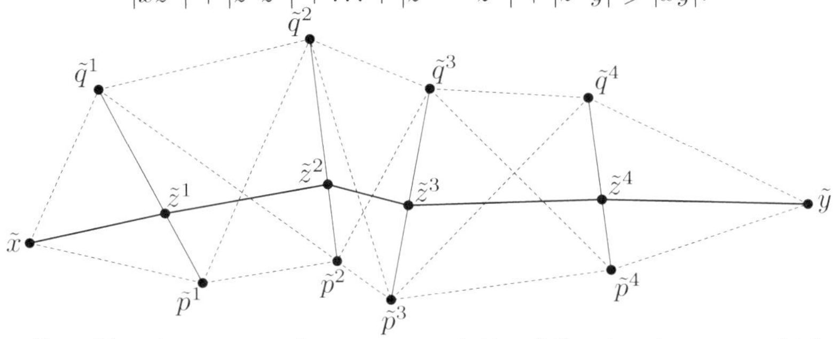

To prove (2n+2)-point comparison, we need the following lemma, which is an easy corollary from Kirszbraun's theorem (5.1).

6.3. Lemma. *Let $\mathcal{L} \in \overline{\mathrm{CBB}}[\kappa]$, $\mathcal{U} \in \overline{\mathrm{CAT}}[\kappa]$, and $Q \subset \mathrm{B}(p, \frac{\varpi^\kappa}{2}) \subset \mathcal{L}$. Then any short map $f\colon Q \to \mathcal{U}$ can be extended to a short map $F\colon \mathcal{L} \to \mathcal{U}$.*

Proof. Directly from Kirszbraun's theorem (5.4 or 5.1), we obtain the case $\kappa \leqslant 0$. Thus it remains to prove the theorem only in the case $\kappa > 0$. After rescaling we may assume that $\kappa = 1$ and therefore $\varpi^\kappa = \pi$.

It is enough to prove that there is a point $z \in \mathcal{U}$ such that $|zf(x)| \leqslant \frac{\pi}{2}$ for all $x \in Q$; once it is proved, the statement follows from Kirszbraun's theorem (5.1).

[6] i.e., perimeter of each triple in p^i, p^{i+1}, q^i and q^{i+1} is $< 2\cdot\pi$ and $|\tilde{p}^i\tilde{q}^i| = |p^iq^i|$, $|\tilde{p}^i\tilde{p}^{i+1}| = |p^ip^{i+1}|$, $|\tilde{q}^i\tilde{q}^{i+1}| = |q^iq^{i+1}|$, $|\tilde{p}^i\tilde{q}^{i+1}| = |p^iq^{i+1}|$ and $|\tilde{p}^{i+1}\ \tilde{q}^i| = |p^{i+1}\ q^i|$.

Further we use the same notations as in the proof of 5.1.

Apply Kirszbraun's theorem (5.4 or 5.1) for $\mathring{f}\colon \mathring{Q} \to \mathring{\mathcal{U}}$ and set $q = \mathring{F}(p)$. Clearly,

$$\langle f(x), q\rangle \geqslant \cos\measuredangle(p, x) > 0$$

for any $x \in Q$. In particular, we have $|q| > 0$. Thus, for $z = \frac{1}{|q|}\cdot q \in \mathcal{U}$, we get $|zf(x)|_{\mathcal{U}} = \measuredangle(z, f(x)) \leqslant \frac{\pi}{2}$ for all $x \in Q$. □

Proof of (2n+2)-point comparison. Direct application of 6.3 gives an array of short maps $f^0, f^1, \dots, f^n\colon \mathbb{M}^3[\kappa] \to \mathcal{U}$ such that

(i) $\tilde{x} \overset{f^0}{\longmapsto} x$, $\tilde{p}^1 \overset{f^0}{\longmapsto} p^1$ and $\tilde{q}^1 \overset{f^0}{\longmapsto} q^1$;

(ii) $\tilde{p}^i \overset{f^i}{\longmapsto} p^i$, $\tilde{q}^i \overset{f^i}{\longmapsto} q^i$ and $\tilde{p}^{i+1} \overset{f^i}{\longmapsto} p^{i+1}$, $\tilde{q}^{i+1} \overset{f^i}{\longmapsto} q^{i+1}$ for $1 \leqslant i \leqslant n-1$;

(iii) $\tilde{p}^n \overset{f^n}{\longmapsto} p^n$, $\tilde{q}^n \overset{f^n}{\longmapsto} q^n$ and $\tilde{y} \overset{f^n}{\longmapsto} y$.

For each $i > 0$, we have that $f^{i-1}|_{[\tilde{p}^i\tilde{q}^i]} = f^i|_{[\tilde{p}^i\tilde{q}^i]}$, since both f^{i-1} and f^i send $[\tilde{p}^i\tilde{q}^i]$ isometrically to a geodesic $[p^iq^i]$ in $\mathcal{U}$ which has to be unique. Thus the curves

$$f^0([\tilde{x}\tilde{z}^1]),\ f^1([\tilde{z}^1\tilde{z}^2]), \dots,\ f^{n-1}([\tilde{z}^{n-1}\tilde{z}^n]),\ f^n([\tilde{z}^n\tilde{y}])$$

can be joined in $\mathcal{U}$ into a curve connecting x to y with length at most

$$|\tilde{x}\tilde{z}^1| + |\tilde{z}^1\tilde{z}^2| + \ldots + |\tilde{z}^{n-1}\ \tilde{z}^n| + |\tilde{z}^n\tilde{y}|.$$ □

7. Comments and open problems

7.1. Open problem. *Find a necessary and sufficient condition for a finite metric space to be isometrically embeddable into some $\overline{\mathrm{CBB}}\lfloor\kappa\rfloor$ space.*

A metric on a finite set $\{a^1, a^2, \dots, a^n\}$, can be described by the matrix with components

$$s^{ij} = |a^ia^j|^2,$$

which we will call the *decrypting matrix*. The set of decrypting matrices of all metrics that admit an isometric embedding into a $\overline{\mathrm{CBB}}\lfloor 0\rfloor$ space form a convex cone, as follows from the fact that the product of $\overline{\mathrm{CBB}}\lfloor 0\rfloor$ spaces is a $\overline{\mathrm{CBB}}\lfloor 0\rfloor$ space. This convexity gives hope that the cone admits an explicit description.

The set of metrics on $\{a^1, a^2, \dots, a^n\}$ that can be embedded into a product of spheres with different radii admits a simple description. Obviously, this gives a sufficient condition for 7.1. This condition is not necessary. For instance, as follows from from a result of Vilms, [16, 2.2], a sufficiently dense finite subset in a generic closed positively curved manifold cannot be embedded into a product of spheres.

Theorem 4.1 gives a necessary condition for 7.1, but the condition is not sufficient. One sees this in the following example constructed by Sergei Ivanov:

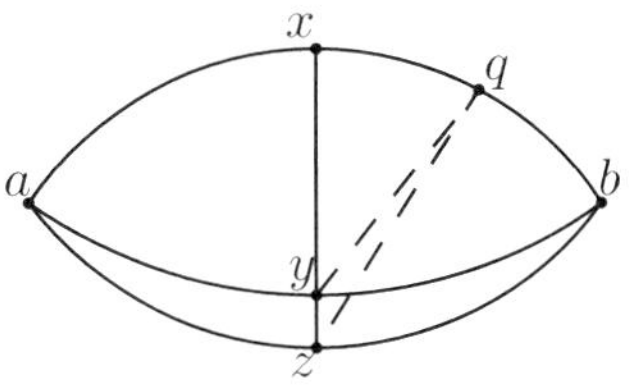

Example. Consider the finite set $\mathcal{F} = \{a, b, x, y, z, q\}$ with distances defined as follows:

(1) $|ab| = 4$;
(2) $|ax| = |ay| = |az| = |bx| = |by| = |bz| = 2$;
(3) $|xy| = 2$, $|yz| = 1$, $|xz| = 3$;
(4) $|xq| = |qb| = 1$ and thus $|aq| = 3$;
(5) $\tilde\measuredangle^0(x\,{}^q_y) = \tilde\measuredangle^0(x\,{}^q_z) = \frac{\pi}{3}$; i.e. $|qy| = \sqrt{3}$ and $|qz| = \sqrt{7}$.

On the diagram the degenerate triangles are marked by solid lines. Note that if one removes from $\mathcal{F}$ the point q then the remaining part can be embedded in a sphere of intrinsic diameter 4 with poles at a and b and the points x, y, z on the equator. On the other hand, if one removes the point a from the space and changes the distance $|zb|$ then it can be isometrically embedded into the plane.

It is straightforward to check that this finite set satisfies the conclusion of Theorem 4.1 for $\kappa = 0$. However, if such a metric appeared as an inherited metric on a subset $\{a, b, x, y, z, q\} \subset \mathcal{L} \in \overline{\mathrm{CBB}}\lfloor 0 \rfloor$ then clearly

$$\measuredangle[x\,{}^a_y] = \measuredangle[y\,{}^a_z] = \measuredangle[y\,{}^b_z] = \tfrac{\pi}{3},$$

contradicting $|bz| = 2$.

The following problem was mentioned by Gromov in [3, 15(b)]

7.2. Open problem. *Describe metrics on an n-point set which are embeddable into* $\mathrm{CAT}\lceil\kappa\rceil$ *spaces.*

The set of metrics on $\{a^1, a^2, \dots, a^n\}$ which can be embedded into a product of trees and hyperbolic spaces admits a simple description using decrypting matrices defined above. Obviously, this gives a sufficient condition for problem 7.2. This condition is not necessary. The existence of a counterexample follows from the same result of Vilms [16, 2.2]; it is sufficient to take a sufficiently dense finite subset in a ball in a generic Hadamard space.

The $(2n+2)$-point comparison (6.2) gives a necessary condition for 7.2 and it is not known whether it is sufficient.

7.3. Open problem. *Let $\mathcal{F}$ be a finite metric space which satisfies $(2n+2)$-point comparison (6.2) for some fixed for κ and any choice of the point array.*

Is it true that $\mathcal{F}$ admits an isometric embedding into a $\mathrm{CAT}\lceil\kappa\rceil$ *space?*

The next conjecture (if true) would give the right generality for Kirszbraun's theorem (5.1). Roughly it states that the example 5.2 is the only obstacle for extending a short map.

7.4. Conjecture. *Assume $\mathcal{L} \in \overline{\mathrm{CBB}}\lfloor 1 \rfloor$, $\mathcal{U} \in \overline{\mathrm{CAT}}\lceil 1 \rceil$, $Q \subset \mathcal{L}$ is a proper subset, and $f\colon Q \to \mathcal{U}$ is a short map that does not admit a short extension to any bigger set $Q' \supset Q$. Then:*

a) *Q is isometric to a sphere in a Hilbert space (of finite or cardinal dimension). Moreover, there is a point $p \in \mathcal{L}$ such that $|pq| = \frac{\pi}{2}$ for any $q \in Q$.*

b) *The map* $f\colon Q\to\mathcal{U}$ *is a global isometric embedding and there is no point* $p'\in\mathcal{U}$ *such that* $|p'q'|=\frac{\pi}{2}$ *for all* $q'\in f(Q)$.

Appendix A. Barycentric simplex

The barycentric simplex was introduced by Kleiner in [7]; it is a construction that works in a general metric space. Roughly, it gives a k-dimensional submanifold for a given "nondegenerate" array of $k+1$ strongly convex functions.

Let us denote by $\Delta^k\subset\mathbb{R}^{k+1}$ the *standard k-simplex*; i.e., $\boldsymbol{x}=(x_0,x_1,\dots,x_n)\in\Delta^k$ if $\sum_{i=0}^k x_i=1$ and $x_i\geqslant 0$ for all i.

Let $\mathcal{X}$ be a metric space and $\boldsymbol{f}=(f^0,f^1,\dots,f^k)\colon\mathcal{X}\to\mathbb{R}^{k+1}$ be a function array. Consider the map $\boldsymbol{f}^\triangle\colon\Delta^k\to\mathcal{X}$, defined by

$$\boldsymbol{f}^\triangle(\boldsymbol{x})=\operatorname{argmin}\sum_{i=0}^k x_i{\cdot}f^i,$$

where argmin f denotes a point of minimum of f. The map $\boldsymbol{f}^\triangle$ will be called a *barycentric simplex* of $\boldsymbol{f}$. In general, a barycentric simplex of a function array might be undefined and need not be unique.

The name comes from the fact that if $\mathcal{X}$ is a Euclidean space and $f^i(x)=\frac{1}{2}{\cdot}|p^ix|^2$ for some array of points $\boldsymbol{p}=(p^0,p^1,\dots,p^k)$, then $\boldsymbol{f}^\triangle(\boldsymbol{x})$ is the barycenter of points p^i with weights x_i.

A barycentric simplex $\boldsymbol{f}^\triangle$ for the function array $f^i(x)=\frac{1}{2}{\cdot}|p^ix|^2$ will also be called a *barycentric simplex with vertices at* $\{p^i\}$.

It is clear from the definition that if $\hat{\boldsymbol{f}}$ is a subarray of $\boldsymbol{f}$, then $\hat{\boldsymbol{f}}^\triangle$ coincides with the restriction of $\boldsymbol{f}^\triangle$ to the corresponding face of Δ^k.

The following theorem shows that the barycentric simplex is defined for an array of strongly convex functions on a complete geodesic space. In order to formulate the theorem, we need to introduce a partial order $\succcurlyeq$ on $\mathbb{R}^{k+1}$.

A.1. Definition. *For two real arrays* $\boldsymbol{v}$, $\boldsymbol{w}\in\mathbb{R}^{k+1}$, $\boldsymbol{v}=(v^0,v^1,\dots,v^k)$ *and* $\boldsymbol{w}=(w^0,w^1,\dots,w^k)$, *we write* $\boldsymbol{v}\succcurlyeq\boldsymbol{w}$ *if* $v^i\geqslant w^i$ *for each* i.

Given a subset $Q\subset\mathbb{R}^{k+1}$, *define its* superset

$$\operatorname{SupSet}Q=\{\boldsymbol{v}\in\mathbb{R}^k\mid\exists\,\boldsymbol{w}\in Q\ \textit{such that}\ \boldsymbol{v}\succcurlyeq\boldsymbol{w}\}.$$

A.2. Theorem on barycentric simplex. *Assume* $\mathcal{X}$ *is a complete geodesic space and* $\boldsymbol{f}=(f^0,f^1,\dots,f^k)\colon\mathcal{X}\to\mathbb{R}^k$ *is an array of strongly convex and locally Lipschitz functions.*

Then the barycentric simplex $\boldsymbol{f}^\triangle\colon\Delta^k\to\mathcal{X}$ *is uniquely defined and moreover:*

a) $\boldsymbol{f}^\triangle$ *is Lipschitz.*

b) *The set* SupSet $\boldsymbol{f}(\mathcal{X})\subset\mathbb{R}^{k+1}$ *is convex, and* $p\in\boldsymbol{f}^\triangle(\Delta^k)$ *if and only if* $\boldsymbol{f}(p)$ *is in* $\partial[\operatorname{SupSet}\boldsymbol{f}(\mathcal{X})]$. *In particular,* $\boldsymbol{f}\circ\boldsymbol{f}^\triangle(\Delta^k)$ *lies on a convex hypersurface in* $\mathbb{R}^{k+1}$.

c) *The restriction* $\boldsymbol{f}|_{\boldsymbol{f}^\triangle(\Delta^k)}$ *has* $C^{\frac{1}{2}}$*-inverse.*

d) *The set* $\mathfrak{S} = \boldsymbol{f}^{\triangle}(\Delta^k)\backslash\boldsymbol{f}^{\triangle}(\partial\Delta^k)$ *is* $C^{\frac{1}{2}}$*-homeomorphic to an open domain in* $\mathbb{R}^k$.

The set $\mathfrak{S}$ described above will be called *Kleiner's spine* of $\boldsymbol{f}$. If $\mathfrak{S}$ is nonempty, we say the barycentric simplex $\boldsymbol{f}^{\triangle}$ is *nondegenerate*.

We precede the proof of the theorem with the following lemma.

A.3. Lemma. *Assume $\mathcal{X}$ is a complete geodesic metric space and let $f\colon \mathcal{X} \to \mathbb{R}$ be a locally Lipschitz, strongly convex function. Then the minimum point $p = \operatorname{argmin} f$ is uniquely defined.*

Proof. Assume that x and y are distinct minimum points of f. Then for the midpoint z of a geodesic $[xy]$ we have

$$f(z) < f(x) = f(y),$$

a contradiction. It only remains to show existence.

Fix a point $p \in \mathcal{X}$; let $\pounds \in \mathbb{R}$ be a Lipschitz constant of f in a neighborhood of p. Without loss of generality, we can assume that f is 1-convex. Consider the function $\varphi(t) = f \circ \operatorname{geod}_{[px]}(t)$. Clearly φ is 1-convex and $\varphi^{+}(0) \geqslant -\pounds$. Setting $\ell = |px|$, we get

$$\begin{aligned} f(x) = \varphi(\ell) &\geqslant \\ &\geqslant f(p) - \pounds\cdot\ell + \tfrac{1}{2}\cdot\ell^2 \geqslant \\ &\geqslant f(p) - \tfrac{1}{2}\cdot\pounds^2. \end{aligned}$$

In particular,

$$s \stackrel{def}{=\!=} \inf\{\, f(x) \mid x \in \mathcal{X} \,\} \geqslant f(p) - \tfrac{1}{2}\cdot\pounds^2.$$

If z is a midpoint of $[xy]$ then

$$s \leqslant f(z) \leqslant \tfrac{1}{2}\cdot f(x) + \tfrac{1}{2}\cdot f(y) - \tfrac{1}{8}\cdot|xy|^2. \qquad ❶$$

Choose a sequence of points $p_n \in \mathcal{X}$ such that $f(p_n) \to s$. Applying ❶, for $x = p_n$, $y = p_m$, we get that (p_n) is a Cauchy sequence. Clearly, $p_n \to \operatorname{argmin} f$. □

Proof of theorem A.2. Without loss of generality, we can assume that each f^i is 1-convex. Thus, for any $\boldsymbol{x} \in \Delta^k$, the convex combination $\sum x_i \cdot f^i\colon \mathcal{X} \to \mathbb{R}$ is also 1-convex. Therefore, according to Lemma A.3, $\boldsymbol{f}^{\triangle}(\boldsymbol{x})$ is defined.

(a). Since Δ^k is compact, it is sufficient to show that $\boldsymbol{f}^{\triangle}$ is locally Lipschitz.

For $\boldsymbol{x}, \boldsymbol{y} \in \Delta^k$, set

$$\begin{aligned} f_{\boldsymbol{x}} &= \textstyle\sum x_i\cdot f^i, & f_{\boldsymbol{y}} &= \textstyle\sum y_i\cdot f^i, \\ p &= \boldsymbol{f}^{\triangle}(\boldsymbol{x}), & q &= \boldsymbol{f}^{\triangle}(\boldsymbol{y}). \end{aligned}$$

Let $\ell = |pq|^2$. Clearly the function $\varphi(t) = f_{\boldsymbol{x}} \circ \mathrm{geod}_{[pq]}(t)$ takes its minimum at 0 and $\psi(t) = f_{\boldsymbol{y}} \circ \mathrm{geod}_{[pq]}(t)$ takes its minimum at ℓ. Thus $\varphi^+(0), \psi^-(\ell) \geqslant 0$[7]. From 1-convexity of $f_{\boldsymbol{y}}$, we have $\psi^+(0) + \psi^-(\ell) + \ell \leqslant 0$.

Let $\pounds$ be a Lipschitz constant for all f^i in a neighborhood $\Omega \ni p$. Then we have $\psi^+(0) \leqslant \varphi^+(0) + \pounds \cdot \|\boldsymbol{x} - \boldsymbol{y}\|_1$, where $\|\boldsymbol{x} - \boldsymbol{y}\|_1 = \sum_{i=0}^k |x_i - y_i|$. That is, given $\boldsymbol{x} \in \Delta^k$, there is a constant $\pounds$ such that

$$|\boldsymbol{f}^\vartriangle(\boldsymbol{x})\ \boldsymbol{f}^\vartriangle(\boldsymbol{y})| = \ell \leqslant \pounds \cdot \|\boldsymbol{x} - \boldsymbol{y}\|_1$$

for any $\boldsymbol{y} \in \Delta^k$. In particular, there is $\varepsilon > 0$ such that if $\|\boldsymbol{x} - \boldsymbol{y}\|_1, \|\boldsymbol{x} - \boldsymbol{z}\|_1 < \varepsilon$, then $\boldsymbol{f}^\vartriangle(\boldsymbol{y}), \boldsymbol{f}^\vartriangle(\boldsymbol{z}) \in \Omega$. Thus, the same argument as above implies

$$|\boldsymbol{f}^\vartriangle(\boldsymbol{y})\ \boldsymbol{f}^\vartriangle(\boldsymbol{z})| = \ell \leqslant \pounds \cdot \|\boldsymbol{y} - \boldsymbol{z}\|_1$$

for any $\boldsymbol{y}$ and $\boldsymbol{z}$ sufficiently close to $\boldsymbol{x}$; i.e. $\boldsymbol{f}^\vartriangle$ is locally Lipschitz.

(b). The "only if" part is trivial, let us prove the "if"-part.

Note that convexity of f^i implies that for any two points $p, q \in \mathcal{X}$ and $t \in [0,1]$ we have

$$(1-t)\cdot \boldsymbol{f}(p) + t \cdot \boldsymbol{f}(q) \succcurlyeq \boldsymbol{f} \circ \mathrm{path}_{[pq]}(t), \qquad ❷$$

where $\mathrm{path}_{[pq]}$ is a geodesic path from p to q; i.e., $\mathrm{path}_{[pq]}(t) = \mathrm{geod}_{[pq]}(\frac{t}{|pq|})$.

From ❷, we have that $\mathrm{SupSet}[\boldsymbol{f}(\mathcal{X})]$ is a convex subset of $\mathbb{R}^{k+1}$. If

$$\max_i \{f^i(q) - f^i(p)\} \geqslant 0$$

for any $q \in \mathcal{X}$, then $\boldsymbol{f}(p)$ lies in the boundary of $\mathrm{SupSet}[\boldsymbol{f}(\mathcal{X})]$. Take a supporting vector $\boldsymbol{x} \in \mathbb{R}^{k+1}$ to $\mathrm{SupSet}[\boldsymbol{f}(\mathcal{X})]$ at $\boldsymbol{f}(p)$. Thus $\boldsymbol{x} \neq \boldsymbol{0}$ and $\sum_i x_i \cdot [w^i - f^i(p)] \geqslant 0$ for any $\boldsymbol{w} \in \mathrm{SupSet}[\boldsymbol{f}(\mathcal{X})]$. In particular, $\sum_i x_i \cdot v_i \geqslant 0$ for any $v = (v_1, \dots, v_k)$ with all $v_i \geqslant 0$. Hence $x_i \geqslant 0$ for all i and $\boldsymbol{x}' = \frac{\boldsymbol{x}}{\|\boldsymbol{x}\|_1} \in \Delta^k$. Thus $p = \boldsymbol{f}^\vartriangle(\boldsymbol{x}')$.

(c). The restriction $\boldsymbol{f}|_{\boldsymbol{f}^\vartriangle(\Delta^k)}$ is Lipschitz. Thus we only have to show that it has a $C^{\frac{1}{2}}$-inverse. Given $\boldsymbol{v} \in \mathbb{R}^{k+1}$, consider the function $h_{\boldsymbol{v}} \colon \mathcal{X} \to \mathbb{R}$ given by

$$h_{\boldsymbol{v}}(p) = \max_i \{f^i(p) - v^i\}.$$

Define a map $\Phi \colon \mathbb{R}^{k+1} \to \mathcal{X}$ by $\Phi(\boldsymbol{v}) = \operatorname{argmin} h_{\boldsymbol{v}}$.

Clearly $h_{\boldsymbol{v}}$ is 1-convex. Thus, according to A.3, $\Phi(\boldsymbol{v})$ is uniquely defined for any $\boldsymbol{v} \in \mathbb{R}^{k+1}$. From (b), for any $p \in \boldsymbol{f}^\vartriangle(\Delta^k)$ we have $\Phi \circ \boldsymbol{f}(p) = p$.

It remains to show that Φ is $C^{\frac{1}{2}}$-continuous. Clearly,

$$|h_{\boldsymbol{v}} - h_{\boldsymbol{w}}| \leqslant \|\boldsymbol{v} - \boldsymbol{w}\|_\infty \overset{def}{=\!=} \max_i \{|v^i - w^i|\},$$

[7]Here $\varphi^\pm$ denotes "signed one sided derivative"; i.e.,

$$\varphi^\pm(t_0) = \lim_{t \to t_0 \pm} \frac{\varphi(t) - \varphi(t_0)}{|t - t_0|}$$

for any $\boldsymbol{v}, \boldsymbol{w} \in \mathbb{R}^{k+1}$. Set $p = \Phi(\boldsymbol{v})$ and $q = \Phi(\boldsymbol{w})$. Since $h_{\boldsymbol{v}}$ and $h_{\boldsymbol{w}}$ are 1-convex,

$$h_{\boldsymbol{v}}(q) \geqslant h_{\boldsymbol{v}}(p) + \tfrac{1}{2}\cdot|pq|^2, \qquad\qquad h_{\boldsymbol{w}}(p) \geqslant h_{\boldsymbol{w}}(q) + \tfrac{1}{2}\cdot|pq|^2.$$

Therefore,

$$|pq|^2 \leqslant 2\cdot\|\boldsymbol{v} - \boldsymbol{w}\|_\infty.$$

Hence the result.

(d). Let $S = \partial\,\mathrm{SupSet}(\boldsymbol{f}(\mathcal{X}))$. Note that orthogonal projection to the hyperplane $\mathbb{W}^k$ in $\mathbb{R}^{k+1}$ defined by equation $x_0 + x_1 + \ldots + x_n = 0$ gives a bi-Lipschitz homeomorphism $S \to \mathbb{W}^k$.

Clearly, $\boldsymbol{f}(\boldsymbol{f}^\triangle(\Delta^k)\backslash\boldsymbol{f}^\triangle(\partial\Delta^k))$ is an open subset of S. Hence the result. □

Appendix B. Helly's theorem

B.1. Helly's theorem. *Let $\mathcal{U} \in \overline{\mathrm{CAT}}[0]$ and $\{K_\alpha\}_{\alpha\in\mathcal{A}}$ be an arbitrary collection of closed bounded convex subsets of $\mathcal{U}$.*

If

$$\bigcap_{\alpha\in\mathcal{A}} K_\alpha = \varnothing,$$

then there is an index array $\alpha_1, \alpha_2, \ldots, \alpha_n \in \mathcal{A}$ such that

$$\bigcap_{i=1}^{n} K_{\alpha_i} = \varnothing.$$

Remarks.

(i) In general, none of K_α might be compact. Thus the the statement is not completely trivial.

(ii) If $\mathcal{U}$ is a Hilbert space (not necessarily separable), then the above result is equivalent to the statement that a convex bounded set which is closed in the ordinary topology forms a compact set in the weak topology.
In fact, one can define the *weak topology* on an arbitrary metric space, by taking exteriors of closed balls as its prebase. Then the result above implies for $\mathcal{U}$ in $\overline{\mathrm{CAT}}[0]$, any closed bounded convex set in $\mathcal{U}$ is compact in the weak topology (this is very similar to the definition given by Monod in [10]).

We present the proof of Lang and Shroeder from [9].

B.2. Lemma. *Let $\mathcal{U} \in \overline{\mathrm{CAT}}[0]$. Given a closed convex set $K \subset \mathcal{U}$ and a point $p \in \mathcal{U}\backslash K$, there is unique point $p^* \in K$ such that $|p^*p| = |Kp|$.*

Proof. Let us first prove uniqueness. Assume there are two points $y', y'' \in K$ so that $|y'p| = |y''p| = |Kp|$. Take z to be the midpoint of $[y'y'']$. Since K is convex, $z \in K$. From comparison, we have that $|zp| < |y'p| = |Kp|$, a contradiction.

The proof of existence is analogous. Take a sequence of points $y_n \in K$ such that $|y_n p| \to |Kp|$. It is enough to show that (y_n) is a Cauchy sequence; thus one could take $p^* = \lim_n y_n$.

Assume (y_n) is not Cauchy, then for some fixed $\varepsilon > 0$, we can choose two subsequences (y'_n) and (y''_n) of (y_n) such that $|y'_n y''_n| \geqslant \varepsilon$ for each n. Set z_n to be the midpoint of $[y'_n y''_n]$; from convexity we have $z_n \in K$. From point-on-side comparison (see page 92), there is $\delta > 0$ such that $|pz_n| \leqslant \max\{|py'_n|, |py''_n|\} - \delta$. Thus

$$\overline{\lim_{n\to\infty}} |pz_n| < |Kx|,$$

a contradiction. □

Proof of B.1. Assume the contrary. Then for any finite set $F \subset \mathcal{A}$,

$$K_F \stackrel{def}{=\!=} \bigcap_{\alpha\in F} K_\alpha \neq \varnothing.$$

We construct a point z such that $z \in K_\alpha$ for each $\alpha \in \mathcal{A}$. Thus we arrive at a contradiction since

$$\bigcap_{\alpha\in\mathcal{A}} K_\alpha = \varnothing.$$

Choose a point $p \in \mathcal{U}$ and set $r = \sup |K_F p|$ where F runs over all finite subsets of $\mathcal{A}$. Let p^*_F be the closest point on K_F from p; according to Lemma B.2, p^*_F exits and is unique.

Take a nested sequence of finite subsets $F_1 \subset F_2 \subset \ldots$ of $\mathcal{A}$, such that $|K_{F_n}\, p| \to r$.

Let us show that $p^*_{F_n}$ is a Cauchy sequence. Indeed, if not then for some fixed $\varepsilon > 0$, we can choose two subsequences (y'_n) and (y''_n) of $(p^*_{F_n})$ such that $|y'_n y''_n| \geqslant \varepsilon$. Set z_n to be midpoint of $[y'_n y''_n]$. From point-on-side comparison (see page 92), there is $\delta > 0$ such that $|pz_n| \leqslant \max\{|py'_n|, |py''_n|\} - \delta$. Thus

$$\overline{\lim_{n\to\infty}} |pz_n| < r.$$

On the other hand, from convexity, each F_n contains all z_k with sufficiently large k, a contradiction.

Thus, $p^*_{F_n}$ converges and we can set $z = \lim_n p^*_{F_n}$. Clearly $|pz| = r$.

Repeat the above arguments for the sequence $F'_n = F_n \cup \{\alpha\}$. As a result, we get another point z' such that $|pz| = |pz'| = r$ and $z, z' \in K_{F_n}$ for all n. Thus, if $z \neq z'$ the midpoint $\hat{z}$ of $[zz']$ would belong to all K_{F_n} and from comparison we would have $|p\hat{z}| < r$, a contradiction.

Thus, $z' = z$; in particular $z \in K_\alpha$ for each $\alpha \in \mathcal{A}$. □

References

[1] S. Alexander, V. Kapovitch and A. Petrunin, *Alexandrov geometry,* a draft is available from `www.math.psu.edu/petrunin`.

[2] D. Burago, Yu. Burago and S. Ivanov, *A course in metric geometry.* Graduate Studies in Mathematics, 33, American Mathematical Society, Providence, RI, 2001.

[3] M. Gromov, CAT(κ)-spaces: Construction and Concentration, *Journal of Mathematical Sciences,* **119** (2004), no. 2, 178–200.

[4] K. Grove and H. Karcher, How to conjugate C^1-close group actions, *Math. Z.* **132** (1973), 11–20.

[5] J. Isbell, Six theorems about metric spaces, *Comment. Math. Helv.* **39** (1964), 65–74.

[6] M. D. Kirszbraun, Über die zusammenziehende und Lipschitzsche Transformationen, *Fundamenta Math.* **22** (1934), 77–108.
[7] B. Kleiner, The local structure of length spaces with curvature bounded above, *Math. Z.* **231** (1999), 409–456.
[8] B. Kleiner and B. Leeb, Rigidity of quasi-isometries for symmetric spaces and Euclidean buildings, *Inst. Hautes Études Sci. Publ. Math.* No. 86 (1997), 115–197 (1998).
[9] U. Lang and V. Schroeder, Kirszbraun's theorem and metric spaces of bounded curvature, *Geom. Funct. Anal.* **7** (1997), no. 3, 535–560.
[10] N. Monod, Superrigidity for irreducible lattices and geometric splitting, *J. Amer. Math. Soc.* **19** (2006), no. 4, 781–814.
[11] C. Plaut, Metric spaces of curvature $\geqslant k$, in *Handbook of geometric topology*, 819–898, North-Holland, Amsterdam, 2002.
[12] Yu. G. Reshetnyak, Nonexpanding maps in a space of curvature no greater than K, *Sibirskii Mat. Zh.* **9** (1968), 918–928 (Russian), English translation: Inextensible mappings in a space of curvature no greater than K, *Siberian Math. Jour.* **9** (1968), 683–689.
[13] K. T. Sturm, Metric spaces of lower bounded curvature, *Exposition. Math.* **17** (1999), no. 1, 35–47.
[14] F. A. Valentine, Contractions in non-Euclidean spaces, *Bull. Amer. Math. Soc.* **50** (1944), 710–713.
[15] F. A. Valentine, A Lipschitz condition preserving extension for a vector function, *Amer. J. Math.* **67** (1945), 83–93.
[16] J. Vilms, Totally geodesic maps, *J. Differential Geometry* **4** (1970), 73–79.

Department of Mathematics, University of Illinois at Urbana-Champaign, 1409 W. Green Street (MC-382), Urbana, Illinois 61801-2975, USA
E-mail address: sba@math.uiuc.edu

Department of Mathematics, University of Toronto, Toronto, Ontario, Canada M5S 2E4
E-mail address: vtk@math.toronto.edu

Mathematics Department, Penn State University, University Park, PA 16802, USA
E-mail address: petrunin@math.psu.edu

Proceedings of 17th Gökova
Geometry-Topology Conference
pp. 110 – 128

On real determinantal quartics

Alex Degtyarev and Ilia Itenberg

ABSTRACT. We describe all possible arrangements of the ten nodes of a generic real determinantal quartic surface in $\mathbb{P}^3$ with nonempty spectrahedral region.

1. Introduction

1.1. Motivation

It is a common understanding that, thanks to the global Torelli theorem for $K3$-surfaces [16] and surjectivity of the period map [13], any reasonable question concerning the topology of singular or real $K3$-surfaces can be reduced to a certain arithmetic problem; many examples, treating the two subjects separately, are found in the literature. However, there are but a few papers where objects that are both real and singular are considered; one can mention [11] and [14], which deal, respectively, with real sextics with a single node in $\mathbb{P}^2$ and real quartic surfaces with a single node in $\mathbb{P}^3$.

In the present paper, we make an attempt to advance this line of research, considering real quartic surfaces with several nodes. Special attention is paid to degenerations of nonsingular quartics, which are used to control the topology of the resulting singular surfaces. Since the classical problem of enumerating all equivariant equisingular deformation types seems rather hopeless (one would expect thousands of classes), we confine ourselves to a very special example arising from convex algebraic geometry. Namely, we describe arrangements of the ten nodes of a generic determinantal quartic with nonempty spectrahedral region, see next subsection for details.

1.2. Principal results

Consider a generic dimension 3 real linear system V of quadrics in $\mathbb{P}^3$. Singular quadrics form a surface $X \subset V \cong \mathbb{P}^3$ of degree 4, which is called a *transversal determinantal quartic* (see Section 4 for details and precise definitions). In other words, we consider a quartic surface $X \subset \mathbb{P}^3$ given by an equation of the form $\det \sum_{i=0}^{3} x_i \bar{q}_i = 0$, where $[x_0 : x_1 : x_2 : x_3]$ are homogeneous coordinates in $\mathbb{P}^3$ and $\bar{q}_0, \bar{q}_1, \bar{q}_2, \bar{q}_3$ are certain fixed

Key words and phrases. Spectrahedron, $K3$-surface, real quartic, singular quartic.
The second author is partially funded by the ANR-09-BLAN-0039-01 grant of *Agence Nationale de la Recherche* and is a member of FRG: Collaborative Research: Mirror Symmetry & Tropical Geometry (Award No. 0854989).

nonzero symmetric (4×4)-matrices. Generically, such a surface is known to have ten nodes.

Whenever present, quadrics given by definite quadratic forms constitute a single connected component of the complement $V_{\mathbb{R}} \smallsetminus X_{\mathbb{R}}$; this component is called the *spectrahedral region* of V. This construction is a special case of a more general framework, see [20] and [17] for details. The study of the shapes of various spectrahedra is a major problem of convex algebraic geometry.

A transversal determinantal quartic has ten nodes, and the original question posed to us by B. Sturmfels was whether all ten can be located in the boundary of the spectrahedral region. (The best known example, constructed explicitly, had eight nodes in the boundary.) We answer this question in the affirmative; moreover, we describe all possible arrangements of the nodes with respect to the components of the complement $V_{\mathbb{R}} \smallsetminus X_{\mathbb{R}}$.

Theorem 1.1. *Let $X \subset \mathbb{P}^3$ be a transversal real determinantal quartic with nonempty spectrahedral region R. Then X has an even number $m \geqslant 0$ of real nodes in the boundary of R and an even number $n \geqslant 0$ of real nodes disjoint from R, so that $2 \leqslant m + n \leqslant 10$. Any pair of even numbers $m, n \geqslant 0$, $2 \leqslant m + n \leqslant 10$, is realized by a quartic as above.*

This theorem is proved in Subsection 5.3.

Remark 1.1. It is worth emphasizing that any transversal real determinantal quartic with nonempty spectrahedral region has at least two real nodes. Note that a similar, and even stronger, statement holds for transversal real determinantal cubics in $\mathbb{P}^3$, which are discriminants of linear systems of plane conics: such a cubic (which is necessarily a Cayley cubic) has nonempty spectrahedral region if and only if at least one of its four nodes is real.

It would be interesting to find explicit matrix representations of transversal real determinantal quartics for all values of m and n allowed by Theorem 1.1.

1.3. Contents of the paper

To prove Theorem 1.1, we analyze the equisingular stratification of the space of complex quartics in $\mathbb{P}^3$, Section 2, and identify the stratum that is formed, up to codimension one subset, by the transversal determinantal quartics, Section 4. Then we describe the sets of cycles that can vanish under certain special nodal degenerations of a real quartic surface, see Section 3. Finally, in Section 5, we show that each transversal real determinantal quartic is obtained by a degeneration of a nonsingular quartic with two nested spheres (the so called *hyperbolic quartic*), and use previously known arithmetical computations in order to construct/prohibit various degenerations of the latter.

2. Singular quartics in $\mathbb{P}^3$

The principal result of this section is Theorem 2.1, which enumerates the equisingular strata of the space of quartics in $\mathbb{P}^3$.

2.1. Integral lattices

A *lattice* is a finitely generated free abelian group L supplied with a symmetric bilinear form $b : L \otimes L \to \mathbb{Z}$. We abbreviate $b(x, y) = x \cdot y$ and $b(x, x) = x^2$. A lattice L is *even* if $x^2 = 0 \bmod 2$ for all $x \in L$. As the transition matrix between two integral bases has determinant ± 1, the determinant $\det L \in \mathbb{Z}$ (*i.e.*, the determinant of the Gram matrix of b in any basis of L) is well defined. A lattice L is called *nondegenerate* if the determinant $\det L \neq 0$; it is called *unimodular* if $\det L = \pm 1$.

Given a lattice L, the bilinear form can be extended to $L \otimes \mathbb{Q}$ by linearity. If L is nondegenerate, the dual group $L^\vee = \operatorname{Hom}(L, \mathbb{Z})$ can be identified with the subgroup

$$\{x \in L \otimes \mathbb{Q} \mid x \cdot y \in \mathbb{Z} \text{ for all } x \in L\}.$$

In particular, L itself can be identified with a subgroup of $L^\vee$.

The group of isometries of a lattice L is denoted by $\operatorname{Aut} L$. Given a vector $a \in L$, $a^2 \neq 0$, the *reflection* against (the hyperplane orthogonal to) a is the automorphism $r_a \colon L \to L$, $x \mapsto x - 2(x \cdot a)a/a^2$, provided that it is well defined, *i.e.*, takes integral vectors to integral vectors. The reflection is always well defined if $a^2 = \pm 1$ or ± 2; if $a^2 = \pm 4$, the reflection r_a is well defined if and only if $a = 0 \bmod 2L^\vee$.

A nondegenerate lattice L is called *elliptic* or *hyperbolic* if its positive inertia index equals 0 or 1, respectively. To any hyperbolic lattice H one can associate a hyperbolic space $\mathbb{P}(C) := C/\mathbb{R}^*$, where $C = C_H := \{x \in H \otimes \mathbb{R} \mid x^2 > 0\}$ is the *positive cone* of H. In particular, given a lattice L and an isometric involution $c \colon L \to L$ with hyperbolic invariant sublattice $L^c_+ = \{x \in L \mid c(x) = x\}$, one can define the space $\mathbb{P}(C^c_+)$. Any subgroup $G \subset \operatorname{Aut} H$ generated by (some) reflections $r_a \colon H \to H$ defined by vectors $a \in H$ with $a^2 < 0$ admits a polyhedral fundamental domain $\mathfrak{P}_G \subset \mathbb{P}(C)$: it is the closure of (any) connected component of the space $\mathbb{P}(C)$ with all mirrors of G removed.

All lattices considered in the paper are even. A *root* in an even lattice is a vector of square (-2). A *root system* is an elliptic lattice generated by roots. We use the standard notation $\mathbf{A}_p$, $p \geqslant 1$, $\mathbf{D}_q$, $q \geqslant 4$, $\mathbf{E}_6$, $\mathbf{E}_7$, $\mathbf{E}_8$ for the irreducible root systems of the same name. Let $\mathbf{U} = \mathbb{Z}u_1 \oplus \mathbb{Z}u_2$, $u_1^2 = u_2^2 = 0$, $u_1 \cdot u_2 = 1$; this lattice is called the *hyperbolic plane*, and any basis (u_1, u_2) as above is called a *standard basis* for $\mathbf{U}$. Given a lattice L and an integer d, the notation $L(d)$ stands for the lattice obtained from L by multiplying the values of the bilinear form by d.

2.2. Singular homological types

Definition 2.1. A *set of* (*simple*) *singularities* is a pair (Σ, σ), where Σ is a root system and σ is a collection of roots of Σ constituting a Weyl chamber of Σ. An isometry $\Sigma_1 \to \Sigma_2$ of two sets of singularities (Σ_i, σ_i), $i = 1, 2$, is *admissible* if it takes σ_1 to σ_2.

Remark 2.1. Any Weyl chamber of a root system Σ can be taken to any other Weyl chamber by an element of the Weyl group of Σ, which extends to any larger lattice containing Σ. For this reason, when speaking about the isomorphism classification of sets of singularities, configurations, and singular homological types (see below), the subset σ in Definition 2.1 can be, and often is disregarded.

Definition 2.2. A *configuration* (extending a given set of singularities (Σ, σ)) is a finite index extension $\tilde{S} \supset S := \Sigma \oplus \mathbb{Z}h$, $h^2 = 4$, satisfying the following conditions:

(1) each root $r \in \tilde{S} \cap (\Sigma \otimes \mathbb{Q})$ belongs to Σ;
(2) $\tilde{S}$ does not contain an element u with $u^2 = 0$ and $u \cdot h = 2$.

An *admissible isometry* of two configurations $\tilde{S}_i \supset S_i = \Sigma_i \oplus \mathbb{Z}h_i$, $i = 1, 2$, is an isometry $\tilde{S}_1 \to \tilde{S}_2$ taking h_1 to h_2 and inducing an admissible isometry $\Sigma_1 \to \Sigma_2$.

Definition 2.3. A *singular homological type* (extending a set of singularities (Σ, σ)) is an extension of the orthogonal direct sum $S := \Sigma \oplus \mathbb{Z}h$, $h^2 = 4$, to a lattice L isomorphic to $2\mathbf{E}_8 \oplus 3\mathbf{U}$, such that the primitive hull $\tilde{S}$ of S in L is a configuration. (The singular homological type is also said to *extend* the configuration $\tilde{S} \supset S$.) An *isomorphism* between two singular homological types $L_i \supset S_i \supset \sigma_i \cup \{h_i\}$, $i = 1, 2$, is an isometry $L_1 \to L_2$ taking h_1 to h_2 and σ_1 to σ_2 (as a set).

A singular homological type is uniquely determined by the collection $\mathcal{H} = (L, h, \sigma)$; then $\Sigma = \Sigma_{\mathcal{H}}$ is the sublattice spanned by σ, and $S = S_{\mathcal{H}} = \Sigma \oplus \mathbb{Z}h$.

Given a singular homological type $\mathcal{H}$, the orthogonal complement $S_{\mathcal{H}}^{\perp}$ is a nondegenerate lattice of positive inertia index 2. Hence, the orthogonal projection of any positive definite 2-subspace $\omega_1 \subset S_{\mathcal{H}}^{\perp} \otimes \mathbb{R}$ to any other such subspace ω_2 is an isomorphism of vector spaces; it can be used to compare orientations of ω_1 and ω_2. Thus, a choice of an orientation of a positive definite 2-subspace in $S_{\mathcal{H}}^{\perp} \otimes \mathbb{R}$ defines a coherent orientation of any other.

Definition 2.4. An *orientation* of a singular homological type $\mathcal{H} = (L, h, \sigma)$ is a choice of coherent orientations of positive definite 2-subspaces of $S_{\mathcal{H}}^{\perp} \otimes \mathbb{R}$. Oriented singular homological types $(\mathcal{H}_i, \mathfrak{o}_i)$, $i = 1, 2$, are *isomorphic* if there is an isomorphism $\mathcal{H}_1 \to \mathcal{H}_2$ taking $\mathfrak{o}_1$ to $\mathfrak{o}_2$. A singular homological type $\mathcal{H}$ is called *symmetric* if $(\mathcal{H}, \mathfrak{o}) \cong (\mathcal{H}, -\mathfrak{o})$, *i.e.*, it $\mathcal{H}$ admits an automorphism reversing orientation.

2.3. Classification of singular quartics

Let $X \subset \mathbb{P}^3$ be a quartic surface with simple singularities only. Denote by $\tilde{X} \to X$ the minimal resolution of singularities of X; it is a minimal $K3$-surface. Introduce the following objects:

- $L_X = H_2(\tilde{X}) = H^2(\tilde{X})$, regarded as a lattice *via* the intersection form (we always identify homology and cohomology *via* the Poincaré duality);
- $\sigma_X \subset L_X$, the set of the classes of the exceptional divisors contracted by the blow-up map $\tilde{X} \to X$;
- $h_X \in L_X$, the class of the pull-back of a generic plane section of X;
- $\omega_X \subset L_X \otimes \mathbb{R}$, the oriented 2-subspace spanned by the real and imaginary parts of the class of a holomorphic 2-form on $\tilde{X}$ (the *period* of $\tilde{X}$).

Note that ω_X is positive definite. According to [18, 21], a triple $\mathcal{H} = (L, h, \sigma)$ has the form (L_X, h_X, σ_X) for a quartic $X \in \mathbb{P}^3$ as above if and only if it is a singular homological type in the sense of Definition 2.3. If this is the case, the above orientation of ω_X defines an orientation of $\mathcal{H}$.

The following theorem is quite expectable; however, we could not find an explicit statement in the literature. The surjectivity part is contained in [21].

Theorem 2.1. *The map sending a quartic surface $X \subset \mathbb{P}^3$ with simple singularities to the pair consisting of its singular homological type $\mathcal{H}_X = (L_X, h_X, \sigma_X)$ and the orientation of the space ω_X establishes a one-to-one correspondence between the set of equisingular deformation classes of quartics with a given set of simple singularities (Σ, σ) and the set of isomorphism classes of oriented abstract singular homological types extending (Σ, σ).*

Proof. Proof of this theorem repeats, almost literally, the proof of a similar theorem for plane sextic curves, see [5]. It is based on Beauville's construction [1] of a fine period space of marked polarized $K3$-surfaces. We omit the details. □

The equisingular stratum of the space of quartic surfaces in $\mathbb{P}^3$ corresponding to an oriented singular homological type $(\mathcal{H}, \mathfrak{o})$ will be denoted by $\mathcal{M}(\mathcal{H}, \mathfrak{o})$. If $\mathcal{H}$ is symmetric, we abbreviate this notation to $\mathcal{M}(\mathcal{H})$. As part of the proof of Theorem 2.1, one obtains an explicit description of the moduli space of quartics, which results in the following formula for its dimension

$$\dim \mathcal{M}(\mathcal{H}, \mathfrak{o})/PGL(4, \mathbb{C}) = 19 - \operatorname{rk} \Sigma_{\mathcal{H}} \tag{1}$$

(similar to the corresponding formula for plane sextics). Note that $\operatorname{rk} \Sigma_{\mathcal{H}} = \#\sigma$ equals the total Milnor number $\mu(X)$ of X.

Remark 2.2. The equisingular deformation classification of quartic surfaces with isolated singularities and at least one non-simple singular point is found in [3, 4]. With a few exceptions, the deformation class of such a quartic is also determined by its (appropriately defined) singular homological type.

3. Real quartics

In this section, we analyze the position of the vanishing cycles of a degeneration of a nonsingular quartic with respect to its period domain. The principal results are Theorems 3.2 and 3.3.

3.1. Real homological types

Given an isometric involution $c\colon L \to L$ on a lattice L, the (± 1)-eigenlattices of c will be denoted by $L^c_\pm = \{x \in L \mid c(x) = \pm x\} \subset L$. If L is nondegenerate, $L^c_\pm$ are the orthogonal complements of each other.

Definition 3.1. A *real homological type* is a triple (L, h, c), where L is a lattice isomorphic to $2\mathbf{E}_8 \oplus 3\mathbf{U}$, $h \in L$ is a vector of square 4, and $c\colon L \to L$ is an isometric involution such that

- the sublattice L^c_+ is hyperbolic, and
- one has $h \in L^c_-$.

An *isomorphism* between two real homological types (L_i, h_i, c_i), $i = 1, 2$, is an isometry $\varphi\colon L_1 \to L_2$ such that $\varphi(h_1) = h_2$ and $\varphi \circ c_1 = c_2 \circ \varphi$.

Definition 3.2. Let (L, h, c) be a real homological type. Consider vectors $e \in L^c_+$ of the following three kinds:

(1) $e^2 = -2$, *i.e.*, e is a root;
(2) $e^2 = -4$ and $e = h \bmod 2L$;
(3) $e^2 = -4$, $e \neq h \bmod 2L$, and e is *decomposable*, *i.e.*, $e = r' - r''$ for a pair of roots $r', r'' \in L$ such that $r' \cdot r'' = r' \cdot h = r'' \cdot h = 0$ and $r'' = -c(r')$.

A *fundamental tower* of (L, h, c) is a triple $\mathfrak{S} \subset \mathfrak{P} \subset \bar{\mathfrak{P}} \subset \mathbb{P}(C^c_+)$ of fundamental domains of the subgroups of $\operatorname{Aut} L^c_+$ generated by the reflections defined by all vectors of L^c_+ of type (1)–(3), (1)–(2), and (1), respectively. (Recall that $C^c_+ \subset L^c_+ \otimes \mathbb{R}$ is the positive cone, see Subsection 2.1.)

Note that, in cases (2) and (3), the conditions imposed imply $e = 0 \bmod 2(L^c_+)^\vee$, *i.e.*, e does define a reflection $r_e\colon L^c_+ \to L^c_+$. Moreover, one can easily see that this reflection extends to an automorphism of the homological type. As a consequence, any two fundamental towers are related by an automorphism of the homological type (in fact, by a sequence of reflections).

Definition 3.3. A real homological type (L, h, c) equipped with a distinguished fundamental tower $\mathfrak{S} \subset \mathfrak{P} \subset \bar{\mathfrak{P}}$ is called a *period lattice*, and the polyhedra $\mathfrak{P}$ and $\bar{\mathfrak{P}}$ are called the *period domains* (more precisely, the period domain of real quartics and that of abstract real $K3$-surfaces, respectively).

As explained in Subsection 2.1, the facets of the polyhedra in Definition 3.2 are (parts of) some of the mirrors (walls) of the respective groups, *i.e.*, hyperplanes orthogonal to vectors of corresponding types. We refer to the type of the vector as the *type* of the corresponding wall.

Remark 3.1. The polyhedra $\mathfrak{P}$ and $\bar{\mathfrak{P}}$ have a certain geometric meaning (see Subsection 3.2 below), whereas $\mathfrak{S}$ does not. However, in many examples, $\mathfrak{S}$ is much easier to compute and, on the other hand, a choice of $\mathfrak{S}$ determines the other two polyhedra: $\mathfrak{P}$ is paved by the copies of $\mathfrak{S}$ obtained from $\mathfrak{S}$ by iterated reflections against (the consecutive images of) the walls of type 3.2(3), and, similarly, $\bar{\mathfrak{P}}$ is paved by the copies of $\mathfrak{P}$ obtained by iterated reflections against the walls of type 3.2(2).

3.2. Invariant periods

A quartic $X \subset \mathbb{P}^3$ is called *real* if it is invariant under the complex conjugation involution $\operatorname{conj}\colon \mathbb{P}^3 \to \mathbb{P}^3$. The involution conj restricts to X and, if X is singular, lifts to the minimal resolution $\tilde{X}$ of singularities of X, turning both into *real K3-surfaces*. A nonsingular real quartic $X \subset \mathbb{P}^3$ gives rise to a real homological type (L_X, h_X, c_X), where $c_X : L_X \to L_X$ is the involution induced by conj.

Two nonsingular real quartics $X, Y \subset \mathbb{P}^3$ are said to be *coarse deformation equivalent* if X is equivariantly deformation equivalent to either Y or the quartic Y' obtained from Y by an orientation reversing automorphism of $\mathbb{P}^3$. A coarse deformation class consists of one or two components of the space of nonsingular real quartics; in the former case, the quartics are called *amphichiral*, in the latter case, *chiral*. The following statement is found in [15].

Theorem 3.1. *Two nonsingular real quartics $X, Y \subset \mathbb{P}^3$ are coarse deformation equivalent if and only if the corresponding real homological types (L_X, h_X, c_X) and (L_Y, h_Y, c_Y) are isomorphic.* □

A complete classification of nonsingular real quartics in $\mathbb{P}^3$ up to equivariant deformation, addressing in particular the chirality problem, and an interpretation of the result in topological terms are found in [12].

Given a real quartic $X \subset \mathbb{P}^3$ or, more generally, a real $K3$-surface (X, conj), a holomorphic 2-form Ω_X on X can be normalized (uniquely up to a nonzero real factor) so that $\mathrm{conj}^* \Omega_X = \bar{\Omega}_X$; such a form is called *real*. The real part $(\omega_X)_+$ of the class of a real form Ω_X belongs to $(L_X)_+^{c_X} \otimes \mathbb{R}$ and defines a point $[(\omega_X)_+]$ in the associated hyperbolic space; this point is called the *invariant period* of X.

Fix a period lattice $(L, h, c; \mathfrak{S} \subset \mathfrak{P} \subset \bar{\mathfrak{P}})$ with the real homological type (L, h, c) isomorphic to that of X. A *marking* of X is a particular choice of an isomorphism $\varphi\colon (L_X, h_X, c_X) \to (L, h, c)$. A marking φ is called *proper* if $\varphi[(\omega_X)_+] \in \mathfrak{P}$. In fact, if X is nonsingular, the image $\varphi[(\omega_X)_+]$ under a proper marking belongs to the interior $\operatorname{Int} \mathfrak{P}$, see, *e.g.*, [15]. It follows that any two proper markings differ by a symmetry of $\mathfrak{P}$.

3.3. Degenerations

A *degeneration* is a smooth family $X_t \subset \mathbb{P}^3$, $t \in [0, 1]$, of real quartics such that all quartics X_t, $t \in (0, 1]$ are nonsingular. For simplicity, we confine ourselves to the case when X_0 has simple nodes only as singularities. Recall that the homology groups $H_2(X_t)$ of the nonsingular members of the family are canonically identified *via* the Gauss-Manin connection, and this common group contains a set of *vanishing cycles* (defined up to sign), one for each node of X_0.

The Gauss-Manin connection can be extended to identify the homology of X_1 with the homology of the minimal resolution $\tilde{X}_0$ of X_0, taking (up to sign) the vanishing cycles to the classes of the exceptional divisors contracted in X_0.

At each real node of X_0, the difference of the local Euler characteristics of the real parts of X_0 and X_1 is ± 1; according to this difference, the node is called *positive* or *negative*, respectively. Negative are the nodes whose vanishing cycles are c_{X_1}-invariant. Below, we are interested in the *non-positive* nodal degenerations, *i.e.*, such that each node of X_0 is either not real or real and negative.

Definition 3.4. Let $(L, h, c; \mathfrak{S} \subset \mathfrak{P} \subset \bar{\mathfrak{P}})$ be a period lattice. A collection of roots $r_i, s'_j, s''_j \in L$, $i = 1, \dots, k$, $j = 1, \dots l$, is called an *admissible system of cycles* if it satisfies the following conditions:

(1) all roots are orthogonal to each other and to h;
(2) the primitive hull in L of the sublattice spanned by r_i, s'_j, s''_j contains no roots other than $\pm r_i$, $\pm s'_j$, $\pm s''_j$, *cf.* 2.2(1);
(3) each root r_i, $i = 1, \dots, k$, belongs to L^c_+ and defines a facet of $\mathfrak{P}$, which is necessarily of type 3.2(1);
(4) for each $j = 1, \dots, l$, one has $c(s'_j) = -s''_j$ and the decomposable invariant vector $s'_j - s''_j$ defines a type 3.2(3) facet of $\mathfrak{S}$.

Theorem 3.2. *Let $(L, h, c; \mathfrak{S} \subset \mathfrak{P} \subset \bar{\mathfrak{P}})$ be a period lattice, and let X_t be a non-positive nodal degeneration with the real homological type of the nonsingular surface $X := X_1$ isomorphic to (L, h, c). Then X admits a proper marking that takes the set of vanishing cycles of X_t to an admissible system of cycles.*

Theorem 3.3. *Given a period lattice $(L, h, c; \mathfrak{S} \subset \mathfrak{P} \subset \bar{\mathfrak{P}})$ and an admissible system of cycles $\sigma = \{r_i, s'_j, s''_j\}$, $i = 1, \dots, k$, $j = 1, \dots, l$, there exists a nonsingular real quartic X and a proper marking $\varphi\colon (L_X, h_X, c_X) \to (L, h, c)$ which identifies σ with the set of vanishing cycles of a certain non-positive nodal degeneration of X.*

Proof of Theorem 3.2. Clearly, the vanishing cycles are orthogonal to each other and to h, *i.e.*, satisfy 3.4(1), as they are geometrically disjoint and can be chosen disjoint from a hyperplane section.

Denote by $r_i \in L_X$, $i = 1, \dots, k$, the vanishing cycles corresponding to the real nodes of X_0, and by $s'_j, s''_j \in L_X$, $j = 1, \dots, l$, those corresponding to the pairs of complex conjugate nodes; the latter are oriented so that $c_X(s'_j) = -s''_j$.

Let $\tilde{X}_0$ be the minimal resolution of singularities of X_0. Recall that, using the Gauss-Manin connection, we identify the homology of X and $\tilde{X}_0$. Let $\tilde{\omega} \in L_X \otimes \mathbb{C}$ be the class realized by a real holomorphic 2-form on $\tilde{X}_0$, and let $[\tilde{\omega}_+] \in \mathfrak{P}_X$ be its invariant part (the invariant period of $\tilde{X}_0$). Note that $[\tilde{\omega}_+]$ does belong to $\mathfrak{P}_X$ as it is the limit of invariant periods of X_t, which are all in $\operatorname{Int} \mathfrak{P}_X$. One has $\operatorname{Pic} \tilde{X}_0 = \tilde{\omega}^\perp \cap L_X$. Denote $\operatorname{Pic}_h \tilde{X}_0 = (h_X)^\perp$, the orthogonal complement of h_X in $\operatorname{Pic} \tilde{X}_0$. Up to sign, any root in $\operatorname{Pic}_h \tilde{X}_0$ is represented by a unique (-2)-curve contracted in X_0, and these are all (-2)-curves contracted. It follows that the roots of $\operatorname{Pic}_h \tilde{X}_0$ are precisely the vanishing cycles; in particular, this implies condition 3.4(2). Thus, the maximal root system in $\operatorname{Pic}_h \tilde{X}_0$ is $(k + 2l)\mathbf{A}_1$, and all its roots define a common face of all its Weyl chambers. Passing to the c_X-invariant part, one easily concludes that the invariant vanishing cycles r_i, $i = 1, \dots, k$, define a common face of all $\mathfrak{P}$-like fundamental polyhedra containing $[\tilde{\omega}_+]$, in particular, of $\mathfrak{P}_X$, whereas the decomposable vectors $s'_j - s''_j$, $j = 1, \dots, l$, define a common face of all $\mathfrak{S}$-like polyhedra containing $[\tilde{\omega}_+]$; for the latter, one can take any polyhedron $\mathfrak{S}'$ containing $[\tilde{\omega}_+]$ and contained in $\mathfrak{P}_X$. Due to Theorem 2.1 and condition 2.2(2) in the definition, one has $s'_j - s''_j \neq h \bmod 2L$; hence the wall defined by this vector is of type 3.2(3).

It remains to consider any proper marking of X_1 and, if necessary, adjust it by a symmetry of $\mathfrak{P}$ to make sure that the polyhedron $\mathfrak{S}'$ constructed above is taken to the preselected polyhedron $\mathfrak{S}$. □

Proof of Theorem 3.3. Let $f_{\mathfrak{P}}$ and $f_{\mathfrak{S}}$ be the intersections of the facets defined in items 3.4(3) and (4), respectively. Notice that $f_{\mathfrak{P}}$ and $f_{\mathfrak{S}}$ are nonempty faces of $\mathfrak{P}$ and $\mathfrak{S}$, respectively (since the facets intersected are mutually orthogonal). One has $f_{\mathfrak{P}} \perp f_{\mathfrak{S}}$ and, since the symmetry about $f_{\mathfrak{S}}$ preserves $\mathfrak{P}$, it also preserves $f_{\mathfrak{P}}$. It follows that the subspace supporting $f_{\mathfrak{S}}$ intersects $f_{\mathfrak{P}}$ at at least one interior point. Let $[\tilde{\omega}_+]$ be such a point, and let $[\tilde{\omega}_-] \subset \mathbb{P}(C^c_{-h})$ be a point in the intersection of the hyperplanes defined by the skew-invariant vectors $s'_j + s''_j$, $j = 1, \ldots, l$, in the hyperbolic space associated with the orthogonal complement L^c_{-h} of h in L^c_-. (Since all hyperplanes are orthogonal to each other, they obviously intersect.) Due to 3.4(2), the pair $([\tilde{\omega}_+], [\tilde{\omega}_-])$ can be chosen generic in the sense that $[\tilde{\omega}_+]$ and $[\tilde{\omega}_-]$ are not simultaneously orthogonal to any root of L which is orthogonal to h and does not belong to $\pm\sigma$.

Let $U \subset \mathbb{P}(C^c_{-h})$ be a sufficiently small neighborhood of $[\tilde{\omega}_-]$. Consider a generic path $([(\omega_+)_t], [(\omega_-)_t]) \in \operatorname{Int} \mathfrak{P} \times U$, $t \in (0, 1]$, converging to the point $([\tilde{\omega}_+], [\tilde{\omega}_-])$. According to [15], it gives rise to a family X_t of properly marked nonsingular real quartics. (Strictly speaking, the path used should avoid a certain codimension 2 subset, see *loc. cit.* for the technical details.) This family can be chosen to converge to a singular quartic X_0 (*cf.*, *e.g.*, [19] for a detailed proof for the similar case of plane sextic, *i.e.*, polarization of square 2), and the limit quartic X_0 is necessarily real. As in the previous proof, considering the Picard group $\operatorname{Pic}_h \tilde{X}_0$ and using the fact that the pair $([\tilde{\omega}_+], [\tilde{\omega}_+])$ is generic, one concludes that the irreducible (-2)-curves contracted in X_0 are precisely those realizing the elements of σ (here, condition 3.4(2) is crucial, which implies that (L, h, σ) is a singular homological type); hence, these elements are the vanishing cycles. □

Remark 3.2. Note that, if negative nodes are present, the real structure does *not* change continuously on the desingularized family $\tilde{X}_t$ of abstract $K3$-surfaces; in fact, the real homological type of the limit surface $\tilde{X}_0$, defined in the obvious way, is not even isomorphic to that of X_t, $t > 0$. However, the real structure does change continuously on the quartics.

4. Complex determinantal quartics

The goal of this section is Theorem 4.8, which identifies the equisingular stratum containing transversal determinantal quartics.

4.1. Notation

Let $\operatorname{Qu}(n) \cong \mathbb{P}^{N(n)}$ be the space of quadrics in $\mathbb{P}^n$; here $N(n) = \frac{1}{2}n(n+3)$. Let, further, $\operatorname{Qu}_r(n) \subset \operatorname{Qu}(n)$, $0 \leqslant r \leqslant n$, be the space of quadrics of corank r. The closure $\Delta(n)$ of $\operatorname{Qu}_1(n)$ is called the *discriminant hypersurface*; it has degree $n+1$.

The singular locus of a quadric Q of corank $r > 0$ is an $(r-1)$-subspace of $\mathbb{P}^n$. Sending Q to its singular locus, one obtains a locally trivial fibration

$$\mathrm{Qu}_r(n) \to \mathrm{Gr}(n+1, r); \tag{2}$$

its fiber is $\mathrm{Qu}_0(n-r)$. (We let $\mathrm{Qu}(0) = \mathrm{Qu}_0(0) = \mathrm{pt.}$) Thus, $\mathrm{Qu}_r(n)$ is a smooth quasi-projective variety and $\dim \mathrm{Qu}_r(n) = N(n) - \frac{1}{2}r(r-1)$.

Definition 4.1. A *geometric hyperplane* is a hyperplane

$$H_p := \{Q \in \mathrm{Qu}(n) \mid Q \ni p\}$$

consisting of all quadrics passing through a fixed point $p \in \mathbb{P}^n$.

For each point $p \in \mathbb{P}^n$, fibration (2) restricts to a locally trivial fibration

$$\mathrm{Qu}_r(n) \smallsetminus H_p \to \mathrm{Gr}(n+1, r) \smallsetminus \mathrm{Gr}(n, r-1) \tag{3}$$

with fiber $\mathrm{Qu}_0(n-r) \smallsetminus H_p$. Here, the difference in the right hand side is the space of all $(r-1)$-planes in $\mathbb{P}^n$ not passing through p.

Since, in this paper, we are mainly concerned with quadrics in $\mathbb{P}^3$, we abbreviate the notation as follows: let $\mathrm{Qu} = \mathrm{Qu}(3)$, $\Delta = \Delta(3)$, and let Δ' and Δ'' be the closures of $\mathrm{Qu}_2(3)$ and $\mathrm{Qu}_3(3)$, respectively. Let also $\Delta^\circ = \mathrm{Qu}_1(3) = \Delta \smallsetminus \Delta'$. One has

$$\dim \mathrm{Qu} = 9, \quad \dim \Delta = \dim \Delta^\circ = 8, \quad \dim \Delta' = 6, \quad \dim \Delta'' = 3.$$

Recall also that $\deg \Delta = 4$ and $\deg \Delta' = 10$ (see [10]).

Let V be a subspace of Qu of dimension 3. Unless $V \in \Delta$, the intersection $\Delta_V := V \cap \Delta$ is a quartic in V. Any quartic $X \in \mathbb{P}^3$ such that the pair $(\mathbb{P}^3, X)$ is isomorphic to (V, Δ_V) above is called a *determinantal quartic*.

A 3-space V is called *transversal* if it is transversal to the strata Δ°, $\Delta' \smallsetminus \Delta''$, and Δ''. Any determinantal quartic $X \subset \mathbb{P}^3$ isomorphic to $\Delta_V \subset V$ is also called *transversal.* If V is transversal, the singular locus $\operatorname{Sing} \Delta_V$ coincides with $V \cap \Delta'$ and consists of ten type $\mathbf{A}_1$ points. Conversely, if $\operatorname{Sing} \Delta_V$ consists of ten type $\mathbf{A}_1$ points, V is transversal.

For a 3-space $V \subset \mathrm{Qu}$ as above, we denote $\Delta_V^\circ = \Delta_V \smallsetminus \Delta'$.

4.2. Some fundamental groups

Observe that the set $\operatorname{Sing} \Delta(n) = \mathrm{Qu}_{\geqslant 2}(n)$ of singular points of $\Delta(n)$ has codimension 3 in $\mathrm{Qu}(n)$. Hence a generic plane section of $\Delta(n)$ is a nonsingular plane curve of degree $n+1$ and, due to Zariski's hyperplane section theorem [22], one has $\pi_1(\mathrm{Qu}_0(n)) = \mathbb{Z}_{n+1}$. A generic plane section of the union $\Delta(n) \cup H_p$ is a transversal union of a nonsingular curve and a line; hence, $\pi_1(\mathrm{Qu}_0(n) \smallsetminus H_p) = \mathbb{Z}$.

Proposition 4.1. *One has* $\pi_1(\Delta^\circ) = 0$.

Proof. The Serre exact sequence of a fibration (2) takes the form

$$\begin{array}{ccccccc} \pi_2(\mathbb{P}^3) & \longrightarrow & \pi_1(\mathrm{Qu}_0(2)) & \longrightarrow & \pi_1(\Delta^\circ) & \longrightarrow & \pi_1(\mathbb{P}^3). \\ \| & & \| & & & & \| \\ \mathbb{Z} & & \mathbb{Z}_3 & & & & 0 \end{array}$$

From this sequence, one concludes that $\pi_1(\Delta^\circ) = H_1(\Delta^\circ)$ is a quotient of $\mathbb{Z}_3$ and, moreover, the inclusion homomorphism $H_1(U \smallsetminus \Delta') \to H_1(\Delta^\circ)$ is onto, where U is a regular neighborhood in Δ of a point $Q \in \Delta'$. A normal 3-plane section of Δ' in Δ is a type $\mathbf{A}_1$ singularity and $U \smallsetminus \Delta'$ is homotopy equivalent to its link. Hence, one has $H_1(U \smallsetminus \Delta') = \mathbb{Z}_2$ and the statement follows. □

Proposition 4.2. *One has $\pi_1(\Delta^\circ \smallsetminus H_p) = \mathbb{Z}_2$ (for any point $p \in \mathbb{P}^3$).*

Proof. Similar to the previous proof, using fibration (3) instead of (2), one concludes that the abelian group $\pi_1(\Delta^\circ \smallsetminus H_p) = H_1(\Delta^\circ \smallsetminus H_p)$ is a quotient of the group $H_1(U \smallsetminus \Delta') = \mathbb{Z}_2$, where U is a regular neighborhood in Δ of a point in Δ'. Thus, it remains to show that $\Delta^\circ \smallsetminus H_p$ admits a nontrivial double covering.

Let $Q \in \Delta^\circ$, $Q \not\ni p$. Denote by q the (only) singular point of Q. Then, there are exactly two planes passing through the line (pq) and tangent to Q along a whole generatrix: the original quadric Q and the two planes are the cones, with the vertex at q, over the section of Q by a generic plane $\alpha \ni p$ and the two tangents to this section passing through p. Clearly, the space of all pairs

$$(Q, \{\text{tangent plane as above}\})$$

is a double covering of $\Delta^\circ \smallsetminus H_p$. This covering is nontrivial: for example, in the family

$$Q_t = \{(x_1 - x_3)^2 + x_1^2 - e^{2\pi it}x_2^2 = 0\}, \qquad t \in [0, 1],$$

the two tangents $x_1 = \pm e^{\pi it}x_2$ are interchanged. (In particular, it follows that the path Q_t, $t \in [0, 1]$, is a non-contractible loop in $\Delta^\circ \smallsetminus H_{(0:0:0:1)}$.) □

Remark 4.1. Similar to Propositions 4.1 and 4.2, one can easily show that all fundamental groups $\pi_1(\mathrm{Qu}_r(n))$ and $\pi_1(\mathrm{Qu}_r(n) \smallsetminus H_p)$ are cyclic.

Corollary 4.3. *Let $p \in \mathbb{P}^3$. Then, for a generic transversal* 3*-plane $V \subset$* Qu, *one has $\pi_1(\Delta^\circ_V) = 0$ and $\pi_1(\Delta^\circ_V \smallsetminus H_p) = \mathbb{Z}_2$.*

Proof. The statement follows from Propositions 4.1 and 4.2 and Zariski type hyperplane section theorem for quasi-projective varieties (see [9, 7, 8] or recent survey [2], Theorem 5.1). □

4.3. The determinantal stratum

In this subsection, we identify the stratum in the space of quartics formed by the transversal determinantal ones.

Lemma 4.4. *Any transversal determinantal quartic has a quadruple of non-coplanar singular points.*

We postpone the proof of this technical statement till Subsection 4.4.

Lemma 4.5. *The space of quintuples* $(V; Q_1, Q_2, Q_3, Q_4)$, *where* $V \subset \mathrm{Qu}$ *is a transversal 3-space and* Q_1, Q_2, Q_3, Q_4 *are four non-coplanar singular points of* Δ_V, *is an irreducible quasi-projective variety of dimension* 24.

Proof. The statement is a tautology, as the 3-subspace $V \subset \mathrm{Qu}$ is uniquely determined by a quadruple of its non-coplanar points Q_1, Q_2, Q_3, Q_4. Thus, the space in question is a Zariski open subset of the irreducible variety $(\Delta' \smallsetminus \Delta'')^4$. □

Since $\dim PGL(4, \mathbb{C}) = 15$, dimension of the moduli space of determinantal quartics is $24 - 15 = 9$, and, comparing the dimensions, see (1), one arrives at the following corollary.

Corollary 4.6. *Transversal determinantal quartics* $X \subset \mathbb{P}^3$ *form a Zariski open subset of a single equisingular stratum of the space of quartics.* □

We denote by $\mathcal{M}_{\mathrm{det}}$ the equisingular stratum containing transversal determinantal quartics. The corresponding configuration and singular homological type are denoted by $\tilde{S}_{\mathrm{det}}$ and $\mathcal{H}_{\mathrm{det}}$, respectively; they extend the set of singularities $\Sigma_{\mathrm{det}} := 10\mathbf{A}_1$. Let $a_1, \ldots, a_{10} \in \Sigma_{\mathrm{det}}$ be generators of the $\mathbf{A}_1$ summands.

Lemma 4.7. *The extension* $\tilde{S}_{\mathrm{det}} \supset S_{\mathrm{det}} := \Sigma_{\mathrm{det}} \oplus \mathbb{Z}h$ *is obtained from* S_{det} *by adjoining the element* $\frac{1}{2}(a_1 + \ldots + a_{10} + h)$. *One has* $\tilde{S}_{\mathrm{det}} \cong \mathbf{U} \oplus \mathbf{E}_8(2) \oplus [-4]$.

Proof. The requirement that $\tilde{S}$ should be an even integral lattice implies that $\tilde{S}_{\mathrm{det}}$ is generated in $S_{\mathrm{det}} \otimes \mathbb{Q}$ by S_{det} and several elements of the form

(1) $\frac{1}{2}(a_1 + \ldots + a_4)$,
(2) $\frac{1}{2}(a_1 + \ldots + a_8)$,
(3) $\frac{1}{2}(a_1 + a_2 + h)$,
(4) $\frac{1}{2}(a_1 + \ldots + a_6 + h)$,
(5) $\frac{1}{2}(a_1 + \ldots + a_{10} + h)$.

(up to reordering of the basis elements a_i).

Case (1) is impossible as the only nontrivial finite index extension of $4\mathbf{A}_1$ is $\mathbf{D}_4$, which contradicts to 2.2(1).

Consider case (2), *i.e.*, assume that $\tilde{S}_{\mathrm{det}}$ contains $a := \frac{1}{2}(a_1 + \ldots + a_8)$. If $\tilde{S}_{\mathrm{det}}$ contained another element a' of the same form, then, up to a further reordering, one would have $a' = \frac{1}{2}(a_1 + \ldots + a_6 + a_9 + a_{10})$, and the difference $a' - a$ would be as in case (1). Hence, a is the only element of $\tilde{S}_{\mathrm{det}} \bmod S_{\mathrm{det}}$ of this form, and each surface in the stratum has eight distinguished singular points. This contradicts Lemma 4.5.

Case (3) contradicts to 2.2(2).

Since cases (1) and (2) have been eliminated, $\tilde{S}_{\mathrm{det}} \bmod S_{\mathrm{det}}$ may contain at most one element as in (4) or (5). In case (4), each surface in the stratum would have six distinguished singular points, which would contradict Lemma 4.5. Thus, either $\tilde{S}_{\mathrm{det}} = S_{\mathrm{det}}$ or

$\tilde{S}_{\rm det} \supset S_{\rm det}$ is the index 2 extension generated by the (only) element (5). Pick a point $p \in \mathbb{P}^3$ and a sufficiently generic transversal quartic Δ_V, so that $H_p \cap \Delta_V$ is nonsingular. Since $\tilde{S}_{\rm det}$ is the primitive hull of $S_{\rm det}$ in $L \cong H_2(\tilde{\Delta}_V)$, from the Poincaré–Lefschetz duality it follows that $H_1(\Delta_V^\circ \smallsetminus H_p) = \operatorname{Ext}(\tilde{S}_{\rm det}/S_{\rm det}, \mathbb{Z})$. Due to Corollary 4.3, one has $[\tilde{S}_{\rm det} : S_{\rm det}] = 2$, and the first statement follows. The isomorphism class of $\tilde{S}_{\rm det}$ is given by a simple computation of the discriminant group and Nikulin's uniqueness theorem [15], Theorem 1.14.2. □

Remark 4.2. Alternatively, case (2) in the proof of Lemma 4.7 can also be eliminated using Corollary 4.3, and case (4) can be eliminated using a refinement of this corollary stating that the group $\pi_1(\Delta_V^\circ \smallsetminus H_p)$ is generated by the group of the link of any of the singular points.

Theorem 4.8. *The configuration $\tilde{S}_{\rm det}$ given by Lemma 4.7 extends to a unique, up to isomorphism, singular homological type $\mathcal{H}_{\rm det}$, which is symmetric. Thus, one has $\mathcal{M}_{\rm det} = \mathcal{M}(\mathcal{H}_{\rm det})$.*

Proof. The uniqueness of a primitive embedding $\tilde{S}_{\rm det} \hookrightarrow L$ of the lattice $\tilde{S}_{\rm det}$ given by Lemma 4.7 follows from [15], Theorem 1.14.4. One has $\tilde{S}_{\rm det}^\perp \cong \mathbf{U} \oplus \mathbf{E}_8(2) \oplus [4]$. Clearly, $\tilde{S}_{\rm det}^\perp$ has a vector of square 2, and the reflection against the hyperplane orthogonal to such a vector is an orientation reversing automorphism. □

Remark 4.3. Alternatively, the fact that $\mathcal{H}_{\rm det}$ is symmetric follows from the obvious existence of real determinantal quartics.

4.4. Proof of Lemma 4.4

We prove a stronger statement: a plane $W \subset V$ cannot contain more than six singular points of a transversal determinantal quartic $\Delta_V \subset V$.

Assume that seven singular points $Q_1, \ldots, Q_7$ of Δ_V belong to a single plane $W \subset V$. Since Δ_V is irreducible, the intersection $\Delta_W := \Delta_V \cap W$ is a curve, which is of degree 4. Furthermore, each point $Q_1, \ldots, Q_7$ is singular for Δ_W.

Since a reduced plane quartic has at most six singular points, Δ_W must have multiple components.

Lemma 4.9. *A pencil $U \subset \operatorname{Qu}$ not contained entirely in Δ' intersects Δ' at at most three points.*

Proof. First, assume that U has a base point singular for all quadrics. Projecting from this point, one obtains a pencil of plane conics, singular conics corresponding to the elements of the intersection $U \cap \Delta'$, and the statement follows from the fact that $\deg \Delta(2) = 3$.

Now, assume that U does not have a singular base point. Let $P_1, P_2 \in U \cap \Delta'$ be two distinct members of U of corank at least 2; they generate U. Since P_1 and P_2 have no common singular points, one has $P_1 = \{x_0x_1 = 0\}$ and $P_2 = \{x_2x_3 = 0\}$ in appropriate homogeneous coordinates, and it is immediate that P_1 and P_2 are the only singular members of U. □

Lemma 4.9 rules out the possibility that Δ_W contains a double line: at most two double lines in Δ_W would contain at most six quadrics in Δ'.

The remaining possibility is that Δ_W is a double conic. In this case Δ_W has no linear components; in particular, no two quadrics $P_1, P_2 \in \Delta_W$ have a common singular point. The quadric Q_1 splits into two distinct planes α_1, α_2. The linear system W restricts to a pencil of conics in α_1 containing at least six distinct singular members, namely, the restrictions of $Q_2, \ldots, Q_7$. (If the restrictions of two distinct quadrics Q_i, Q_j coincided, Q_i and Q_j would have a common singular point.) Since $\deg \Delta(2) = 3 < 6$, all members of the restricted pencil are singular. Hence, the vertex of each quadric $P \in \Delta_W \smallsetminus \Delta'$ belongs to α_1. The same argument shows that the vertex of P also belongs to α_2, *i.e.*, P and Q_1 have a common singular point. This contradiction concludes the proof of Lemma 4.4. □

Remark 4.4. Using the surjectivity of the period map and the Riemann–Roch theorem for $K3$-surfaces, one can easily show that the stratum $\mathcal{M}_{\mathrm{det}}$ does contain a quartic with all ten singular points coplanar (lying in a curve of degree two). In particular, determinantal quartics form a proper subset of $\mathcal{M}_{\mathrm{det}}$.

5. Real determinantal quartics

In this section, we discuss the topology of a determinantal quartic with nonempty spectrahedral domain and prove Theorem 1.1.

5.1. Geometric real structures

We always consider the space $\mathrm{Qu}(n)$ with its *geometric real structure*, *i.e.*, the one induced by the complex conjugation $\mathrm{conj}\colon \mathbb{P}^n \to \mathbb{P}^n$. All real quadratic forms constitute a linear space $\mathbb{R}^{N(n)+1}$, and one has a double covering $S^{N(n)} \to \mathrm{Qu}(n)_{\mathbb{R}}$, where $S^{N(n)}$ is the unit sphere in $\mathbb{R}^{N(n)+1}$.

We reserve the notation $\bar{\ }$ for the lift from $\mathrm{Qu}(n)_{\mathbb{R}}$ to $S^{N(n)}$; in particular, one has real discriminant hypersurfaces $\bar{\Delta}(n) \subset S^{N(n)}$ and $\bar{\Delta} \subset S^9$.

Recall that a real quadratic form $\bar{q}$ has a well defined *index* $\operatorname{ind} \bar{q}$ (the negative inertia index of $\bar{q}$); one has $0 \leqslant \operatorname{ind} \bar{q} \leqslant n+1$.

A *real determinantal quartic* is a real quartic $X \subset \mathbb{P}^3$ equivariantly isomorphic to a quartic $\Delta_V \subset V$, where $V \subset \mathrm{Qu}$ is a 3-subspace real with respect to the geometric real structure. Given such a quartic X, the *spectrahedral region* of X is the (only) connected component of the complement $\mathbb{P}^3_{\mathbb{R}} \smallsetminus X_{\mathbb{R}}$ constituted by the quadrics represented by quadratic forms of index 0 (equivalently, those of maximal index 4).

Lemma 5.1. *Let $X \subset \mathbb{P}^3$ be a real determinantal quartic. Then any real line meeting the spectrahedral region of X intersects X at four real points (counted with multiplicities). In other words, all intersection points are real.*

Proof. Identify $(\mathbb{P}^3, X)$ with a real pair (V, Δ_V) and let W be the image of the line in question. Consider the lift $\bar{W} \subset S^9$. The index function $\operatorname{ind}\colon \bar{W} \to \mathbb{Z}$ is locally constant on $\bar{W} \smallsetminus \bar{\Delta}$, and its increment δ_p at an intersection point $p \in \bar{W} \cap \bar{\Delta}$ of multiplicity m_p is subject to the conditions $|\delta_p| \leqslant m_p$, $\delta_p = m_p \bmod 2$. Any point Q in the spectrahedral region of Δ_V lifts to a pair $\bar{q}$, $-\bar{q}$ of quadratic forms, so that one has $\operatorname{ind} \bar{q} = 4$, $\operatorname{ind}(-\bar{q}) = 0$. Since the two indices differ by 4, the intersection $\bar{W} \cap \bar{\Delta}$ must consist of at least eight points (counted with multiplicities). □

Corollary 5.2. *Let $X \subset \mathbb{P}^3$ be a real determinantal quartic with nonempty spectrahedral region. Then X is a non-positive degeneration of a nonsingular real quartic $X' \subset \mathbb{P}^3$ with the real part $X'_{\mathbb{R}}$ constituted by a nested pair of spheres.* □

5.2. Preliminary computation

Consider the real homological type (L, h, c) corresponding to nonsingular real quartics with two nested spheres; such quartics are amphichiral. According to [15], one has

$$L_+^c \cong \mathbf{E}_8(2) \oplus 2\mathbf{A}_1 \oplus \mathbf{U}, \qquad L_-^c \cong \mathbf{E}_8(2) \oplus 2\mathbf{A}_1(-1). \tag{4}$$

Fix standard bases $e_1, \ldots, e_8$, v_1, v_2, and u_1, u_2 for $\mathbf{E}_8(2)$, $2\mathbf{A}_1$, and $\mathbf{U}$, respectively, and let $e'_1, \ldots, e'_8$, v'_1, v'_2 be the 'matching' standard basis for L_-^c, so that the sum $r + r'$ of two basis vectors $r \in L_+^c$, $r' \in L_-^c$ of the same name is divisible by 2 in L. In view of [15],

$$h = v'_1 + v'_2, \qquad h = v_1 + v_2 \bmod 2L. \tag{5}$$

Let $\mathfrak{S} \subset \mathfrak{P} \subset \bar{\mathfrak{P}}$ be a fundamental tower of (L, h, c). The polyhedron $\mathfrak{S}$ is finite; its Coxeter scheme, computed in [14], is shown in Figure 1, and $\mathfrak{S}$ can be chosen to be bounded by the walls orthogonal to the vectors indicated in the figure. (In the figure, walls of type 3.2(1) and (3) are represented by ∘ and •, respectively, and the only wall of type 3.2(2) is represented by a circled bullet. Whenever the hyperplanes supporting two walls intersect at an angle π/n, $n \geqslant 2$, the corresponding vertices are connected by $(n-2)$ edges. Note that, in the case under consideration, any two walls do intersect.)

Clearly, $\mathfrak{P}$ is paved by the (infinitely many) copies of $\mathfrak{S}$ obtained by iterated reflections against the walls of type 3.2(3) (vertices $e_1, \ldots, e_8$, e_{12}, and e_{13} in the figure), and $\bar{\mathfrak{P}}$ is the union of $\mathfrak{P}$ and its image under the reflection against the only wall e_9 of type 3.2(2); see Definition 3.2 and Remark 3.1.

Let $X \subset \mathbb{P}^3$ be a properly marked nonsingular quartic of type (L, h, c).

Lemma 5.3. *The classes realized in L_+^c by the inner and outer spheres of $X_{\mathbb{R}}$ are* $\mathrm{sp}_{in} = e_{11}$ *and* $\mathrm{sp}_{out} = e_{11} + e_9$, *respectively.*

Proof. Let G be the graph obtained from the Coxeter scheme of $\bar{\mathfrak{P}}$ by removing all but simple edges. According to [6], Theorem 16.1.1, any vertex of G of valency > 2 is a class realized by a spherical component of $X_{\mathbb{R}}$. Clearly, e_{11} and $e_{11} + e_9$ (obtained from e_{11} by reflection) are two such vertices. The outer sphere is definitely not contractible. Hence, the vector e_{11}, which defines a wall of $\mathfrak{P}$ and thus can serve as a vanishing cycle, see Theorem 3.3, represents the inner sphere. □

Denote by $L_+^0 \subset L_+^c$ the sublattice spanned by $e_1, \ldots, e_8$, e_{12}, and e_{13}.

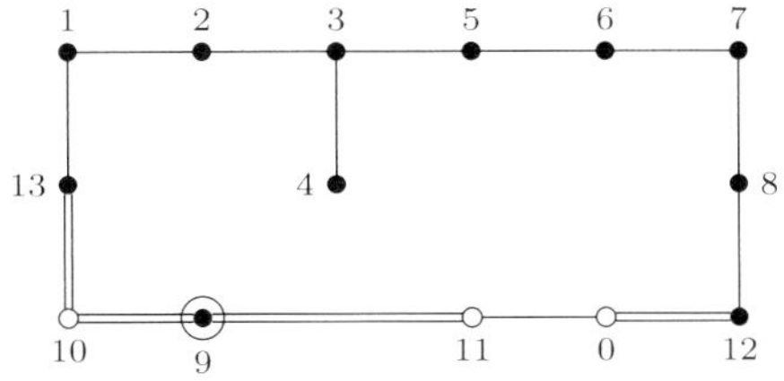

$$e_0 = u_1 - u_2, \qquad e_{11} = u_2 - v_1,$$
$$e_9 = v_1 - v_2, \qquad e_{12} = 2u_2 + e_8^*,$$
$$e_{10} = v_2, \qquad e_{13} = 2(u_1 + u_2) - v_1 - v_2 + e_1^*,$$

$$e_1^* = -4e_1 - 7e_2 - 10e_3 - 5e_4 - 8e_5 - 6e_6 - 4e_7 - 2e_8,$$
$$e_8^* = -2e_1 - 4e_2 - 6e_3 - 3e_4 - 5e_5 - 4e_6 - 3e_7 - 2e_8.$$

FIGURE 1. The fundamental polyhedron $\mathfrak{S}$

Lemma 5.4. *Each real vanishing cycle $r \in L_+^c$ of a non-positive nodal degeneration of X is of one of the following three forms:*

(1) $e_{11} = \mathrm{sp}_{in}$ *(the inner sphere shrinks to a point);*
(2) $e_0 + d$, $d \in L_+^0$ *(a common point of the two spheres);*
(3) $e_{10} + d$, $d \in L_+^0$ *(a node in the outer sphere).*

For each pair s', s'' of conjugate vanishing cycles, the invariant decomposable vector $s' - s''$ belongs to L_+^0.

Proof. Each real vanishing cycle is a wall of $\mathfrak{P}$ of type 3.2(1), see Theorem 3.2. From the description of $\mathfrak{P}$ in terms of $\mathfrak{S}$ and Figure 1 it follows that any such wall either is e_{11} or is obtained from e_0 or e_{10} by iterated reflections against the walls $e_1, \ldots, e_8$, e_{12}, and e_{13} (and their consecutive images). The geometry of the corresponding degeneration is easily seen from comparing the vanishing cycle r against the classes of the spheres: one has $r = \mathrm{sp}_{\mathrm{in}}$ in case (1), $r \cdot \mathrm{sp}_{\mathrm{in}} = r \cdot \mathrm{sp}_{\mathrm{out}} = 1$ in case (2), and $r \cdot \mathrm{sp}_{\mathrm{in}} = 0$, $r \cdot \mathrm{sp}_{\mathrm{out}} = 2$ in case (3).

For a pair s', s'' of conjugate vanishing cycles, the vector $s' - s''$ is either one of the type 3.2(3) walls of $\mathfrak{S}$ or one of their consecutive images under reflections, see Theorem 3.2. □

Lemma 5.5. *The lattice L_-^c does not contain a quintuple t_i, $i = 1, \ldots, 5$, of pairwise orthogonal vectors of square (-4) such that $t_1 + \ldots + t_5 = h \bmod 2L_-^c$.*

Proof. In view of (4) and (5), the vector h is characteristic, *i.e.*, $a^2 + a \cdot h = 0 \bmod 4$ for any $a \in L_-^c$. Hence, $(h + 2a)^2 = 4 \bmod 16$ for any $a \in L_-^c$. On the other hand, $(t_1 + \ldots + t_5)^2 = -20 = -4 \bmod 16$. □

5.3. Proof of Theorem 1.1

We keep the notation of Subsection 5.2. The assumption that the spectrahedral region R of X is nonempty rules out real vanishing cycles of type 5.4(1). Assume that X has m vanishing cycles of type 5.4(2) and n vanishing cycles of type (3) (and $10-m-n$ imaginary vanishing cycles split into conjugate pairs). Using Lemma 5.4 and the description of the vectors involved given in Figure 1, one can easily see that the parities of the coefficients of v_1 and v_2 in the sum of all ten vanishing cycles differ by $n \bmod 2$. Due to (5) and Lemma 4.7, n is even, and so is m.

If $m = n = 0$, then X has five pairs of complex conjugate vanishing cycles s'_j, $s''_j = -c(s'_j)$, $j = 1, \ldots, 5$, and the skew-invariant vectors $s'_j + s''_j \in L^c_-$ form a quintuple contradicting Lemma 5.5.

For the construction, relabel the nine vertices of type 3.2(3) in the edges of the Coxeter scheme consecutively, *i.e.*, let $e_{13} = w_1$, $e_i = w_{i+1}$ for $i = 1, 2, 3$, $e_i = w_i$ for $i = 5, \ldots, 8$, and $e_{12} = w_9$. Pick a pair m, n of even integers as in the statement, let p be $5 - \frac{1}{2}(m+n)$, and consider the following vectors:

$$\begin{aligned} r'_i &= e_0 + w_9 + \ldots + w_{11-i}, && i = 1, \ldots, m \text{ (if } m > 0),\\ r''_j &= e_{10} + w_1 + \ldots + w_{j-1}, && j = 1, \ldots, n \text{ (if } n > 0),\\ t_k &= w_{n+2k-1}, && k = 1, \ldots, p \text{ (if } p > 0). \end{aligned}$$

It is straightforward to check that:

(1) each r'_i is obtained by a sequence of reflections from the vertex e_0, *i.e.*, is as in Lemma 5.4(2);
(2) each r''_j is obtained by a sequence of reflections from the vertex e_{10}, *i.e.*, is as in Lemma 5.4(3);
(3) each t_k is a wall of $\mathfrak{S}$ of type 3.2(3);
(4) all vectors are pairwise orthogonal;
(5) all vectors are linearly independent in $L^c_+/2L^c_+$;
(6) the sum of all ten vectors equals $h \bmod 2L$.

Furthermore, assuming that $p \leqslant 4$, one can easily find pairwise orthogonal vectors $t'_1, \ldots, t'_p \in L^c_-$ such that $t'^2_k = -4$, $t'_k \cdot h = 0$, and $t'_k = t_k \bmod 2L$, $k = 1, \ldots, p$. Indeed, consider the vectors $w'_1 = v'_1 - v'_2 + e'^*_1$, $w'_3 = e'_2$, $w'_5 = e'_5$, $w'_7 = e'_7$, $w'_9 = e'^*_8$. They are orthogonal to h and have square (-4), and any sequence of up to four consecutive vectors is orthogonal. (In fact, all five vectors are pairwise orthogonal except that $w'_1 \cdot w'_9 \neq 0$.) Now, one can take for t'_k the 'matching' vectors w'_*.

Finally, the set σ constituted by the ten vectors

$$r'_1, \ldots, r'_m,\ r''_1, \ldots, r''_n,\ \tfrac{1}{2}(t'_1 \pm t_1), \ldots, \tfrac{1}{2}(t'_p \pm t_p)$$

is an admissible system of cycles, see Definition 3.4; due to Theorem 3.3, it can serve as the set of vanishing cycles of a non-positive nodal degeneration of X. On the other hand, the set σ satisfies Lemma 4.7; hence, according to Corollary 4.6 and Theorem 4.8, a generic degeneration of X contracting these vanishing cycles is a transversal determinantal quartic. □

Acknowledgements: We are grateful to B. Sturmfels for attracting our attention to the problem and for motivating discussions. This paper was conceived during the second author's stay at Mathematical Sciences Research Institute at Berkeley and completed during the first author's visit to Université de Strasbourg.

References

[1] A. Beauville, Application aux espaces de modules, in: Géométrie des surfaces $K3$: modules et périodes, *Astérisque*, **126** (1985), 141–152.

[2] D. Chéniot, Homotopical variation, in: *Singularities II*, Contemp. Math. **475** Amer. Math. Soc., Providence, RI (2008), 11–41.

[3] A. Degtyarev, Classification of surfaces of degree four having a non-simple singular point, *Izv. Akad. Nauk SSSR, Ser. mat.*, **53** (1989), no. 6, 1269–1290 (Russian). English transl. in *Math. USSR-Izv.*, **35** (1990), no. 3, 607–627.

[4] A. Degtyarev, Classification of quartic surfaces that have a nonsimple singular point. II, in: *Geom. i Topol. 1*, Zap. Nauchn. Sem. Leningrad. Otdel. Mat. Inst. Steklov. (LOMI) **193** (1991), 10–38, 161 (Russian). English transl. in: *Topology of manifolds and varieties*, Adv. Soviet Math. **18** Amer. Math. Soc., Providence, RI (1994), 23–54.

[5] A. Degtyarev, On deformations of singular plane sextics, *J. Algebraic Geom.*, **17** (2008), 101–135.

[6] A. Degtyarev, I. Itenberg and V. Kharlamov, *Real Enriques surfaces*, Lecture Notes in Math. **1746** Springer–Verlag, 2000.

[7] M. Goresky and R. MacPherson, Stratified Morse theory, in: *Singularities, Part 1*, (Arcata, Calif., 1981), Proc. Sympos. Pure Math. **40**, Amer. Math. Soc., Providence, R.I. (1983), 517–533.

[8] M. Goresky and R. MacPherson, *Stratified Morse theory*, Ergebnisse der Mathematik und ihrer Grenzgebiete (3) **14** Springer–Verlag, Berlin 1988.

[9] H. A. Hamm and Lê Dung Tráng, Lefschetz theorems on quasiprojective varieties, *Bull. Soc. Math. France*, **113** (1985), no. 2, 123–142.

[10] J. Harris and L. W. Tu, On symmetric and skew-symmetric determinantal varieties, *Topology*, **23** (1984), 71–84.

[11] I. Itenberg, Curves of degree 6 with one non-degenerate double point and groups of monodromy of non-singular curves, *Lecture Notes in Mathematics*, **1524**, *Real Algebraic Geometry, Proceedings, Rennes 1991*, (1992), 267–288.

[12] V. Kharlamov, On classification of nonsingular surfaces of degree 4 in $\mathbb{R}P^3$ with respect to rigid isotopies, *Funkt. Anal. i Priloz.*, (1984), no. 1, 49–56 (Russian). English transl. in *Functional Anal. Appl.*, **18** (1984), no. 1, 39–45.

[13] V. Kulikov, Surjectivity of the period mapping for $K3$-surfaces, *Uspekhi Mat. Nauk*, **32** (1977), no. 4, 257–258.

[14] S. Moriceau, *Surfaces de degré* 4 *avec un point double non dégénéré dans l'espace projectif réel de dimension* 3, Ph. D. Thesis Université de Rennes I (2004).

[15] V. V. Nikulin, Integer quadratic forms and some of their geometrical applications, *Izv. Akad. Nauk SSSR, Ser. Mat*, **43** (1979), 111–177 (Russian). English transl. in *Math. USSR–Izv.*, **14** (1980), 103–167.

[16] I. Piatetski-Shapiro and I. Shafarevich, Torelli's theorem for algebraic surfaces of type $K3$, *Izv. Akad. Nauk SSSR*, **35** (1971), 530–572 (Russian). English transl. in *Math. USSR–Izv.*, **5** 547–588.

[17] Ph. Rostalski and B. Sturmfels, Dualities in convex algebraic geometry, arXiv preprint, arXiv:1006.4894.

[18] B. Saint-Donat, Projective models of $K-3$ surfaces, *Amer. J. Math.*, **96** (1974), 602–639.

[19] I. Shimada, Lattice Zariski k-ples of plane sextic curves and Z-splitting curves for double plane sextics, *Michigan Math. J.*, **59** (2010), no. 3, 621–665.
[20] R. Sanyal, F. Sottile and B. Sturmfels, Orbitopes, arXiv preprint, arXiv:0911.5436.
[21] T. Urabe, Elementary transformations of Dynkin graphs and singularities on quartic surfaces, *Invent. Math.*, **87** (1987), no. 3, 549–572.
[22] O. Zariski, A theorem on the Poincaré group of an algebraic hypersurface, *Ann. of Math. (2)*, **38** (1937), no. 1, 131–141.

Bilkent University, Department of Mathematics, 06800 Ankara, Turkey
E-mail address: degt@fen.bilkent.edu.tr

Université de Strasbourg, IRMA and Institut Universitaire de France, 7 rue René Descartes, 67084 Strasbourg Cedex, France
E-mail address: ilia.itenberg@math.unistra.fr

Proceedings of 17th Gökova
Geometry-Topology Conference
pp. 129 – 134

On the curvature of the real amoeba

Mikael Passare and Jean-Jacques Risler

Abstract. For a real smooth algebraic curve $A \subset (\mathbb{C}^*)^2$, the amoeba $\mathcal{A} \subset \mathbb{R}^2$ is the image of A under the map Log : $(x, y) \mapsto (\log |x|, \log |y|)$. We describe a universal bound for the total curvature of the real amoeba $\mathcal{A}_{\mathbb{R}A}$ and we prove that this bound is reached if and only if the curve A is a simple Harnack curve in the sense of Mikhalkin.

1. Introduction

Let $A \subset (\mathbb{C}^*)^2$ be a smooth real algebraic curve with a non-degenerated Newton polygon Δ defined by an equation $f = 0$, where $f : \mathbb{C}^2 \to \mathbb{C}$ is a reduced polynomial with real coefficients. Let $\mathbb{R}A \subset (\mathbb{R}^*)^2$ stand for the real part of A and Log be the map $Log : (\mathbb{C}^*)^2 \to \mathbb{R}^2$, $(z_1, z_2) \mapsto (\log |z_1|, \log |z_2|)$, $L = Log_{|A}$ the restriction of Log to the curve A, $\mathcal{A}_A = L(A)$, $\mathcal{A}_{\mathbb{R}A} = L(\mathbb{R}A)$, $F \subset A$ the critical set of L. The set $\mathcal{A}_A$ is called the *amoeba* of A, and $\mathcal{A}_{\mathbb{R}A}$ its *real amoeba* (see [1]). Notice that $\mathcal{A}_{\mathbb{R}A}$ is a real curve since for each quadrant $Q \subset (\mathbb{R}^*)^2$, $Log_{|Q}$ is a diffeomorphism.

G. Mikhalkin ([5]) defined “simple Harnack curves” (see below), and used the notion of amoebas to prove the uniqueness of the topological type of the pair $(\mathbb{R}\bar{A}, T_\Delta)$ when A is a simple Harnack curve and $\bar{A}$ its closure in the toric surface T_Δ. In the paper [8], it is proved that $\text{Area}(\mathcal{A}) \leq \pi^2 \text{Area}(\Delta)$ and in [7] the auhors prove A is a simple Harnack curve if and only if there is equality above, i.e., if and only if the Area of $\mathcal{A}$ is maximal.

In this paper, we prove a similar result, but for the total curvature of the real amoeba (instead of the volume of the amoeba); notice that for a hypersurface (in higher dimension), the total curvature of the real amoeba is always finite, but not the volume of the amoeba which is not bounded in general. We then hope to generalize the results of this paper in higher dimension (at least for surfaces).

2. Amoebas and curvature

Let T_Δ be the toric surface associated to Δ, $\tilde{l}_i$ $(1 \leq i \leq n)$ the sides of Δ, d_i the integer length of $\tilde{l}_i$, $\bar{A}$ (resp. $\mathbb{R}\bar{A}$) the closure of the image of A in T_Δ (resp. in $\mathbb{R}T_\Delta$). We denote by l_i the component of the divisor $T_\Delta^\infty := T_\Delta \setminus ((\mathbb{C}^*)^2)$ corresponding to the side $\tilde{l}_i \in \partial\Delta$. We assume that $\bar{A}$ is smooth and transverse to T_Δ^∞.

1991 *Mathematics Subject Classification.* 14H20, 14H50, 14P59.
Key words and phrases. Real plane algebraic curves, Curvature.

Set γ for the logarithmic Gauss map $\gamma : A \to \mathbb{C}P^1$ (see [5]); the map γ is defined by $\gamma(z_1, z_2) = [z_1 \partial f/\partial z_1, z_2 \partial f/\partial z_2]$. The real Logarithmic Gauss map $\gamma_{\mathbb{R}} : \mathbb{R}A \to \mathbb{R}P^1$ is the restriction of γ to $\mathbb{R}A$. We have $F = \gamma^{-1}(\mathbb{R}P^1)$ ([5], Lemma 3), and the following commutative diagram:

$$\begin{array}{ccc} \mathbb{R}A & & \\ {\scriptstyle L}\downarrow & \searrow^{\gamma_{\mathbb{R}}} & \\ \mathcal{A}_{\mathbb{R}A} & \xrightarrow[g]{} & \mathbb{R}P^1 \end{array}$$

where g is the usual Gauss map, defined on the smooth part of $\mathcal{A}_{\mathbb{R}A}$. The logarithmic Gauss map $\gamma : A \to \mathbb{C}P^1$ extends to a map $\bar{\gamma} : \bar{A} \to \mathbb{C}P^1$ of degree $2vol(\Delta)$ ([5]), therefore the map $\gamma_{\mathbb{R}}$ has finite fibers of cardinality $\leq 2vol(\Delta)$ which implies that the fibers of g are also of cardinality $\leq 2vol(\Delta)$.

If k denotes the curvature function on the curve $\mathcal{A}_{\mathbb{R}A}$ (for any orientation), we have

$$\int_{\mathcal{A}_{\mathbb{R}A}} |k| \leq 2\pi vol(\Delta) \tag{1}$$

(see [9] or [4]), since $\mathrm{vol}(\mathbb{R}P^1) = \pi$ and because $\int_{\mathcal{A}_{\mathbb{R}A}} |k|$ is the volume of $\mathrm{im}(g) \subset \mathbb{R}P^1$ (counted with multiplicities, the multiplicity of $x \in \mathbb{R}P^1$ being the cardinality of the fiber $g^{-1}(x)$ which is $\leq 2vol(\Delta)$).

Definition 2.1. We say that the real amoeba $\mathcal{A}_{\mathbb{R}A}$ has maximal curvature if there is equality in (1).

Lemma 2.1. *Let $A \subset (\mathbb{C}^*)^2$ be a real smooth algebraic curve. Then the following conditions are equivalent:*

(1) *$\mathcal{A}_{\mathbb{R}A}$ has maximal curvature*
(2) *The logarithmic Gauss map $\gamma : A \to \mathbb{C}P^1$ is totally real (i.e., $\gamma^{-1}(x) \subset \mathbb{R}A$ for all $x \in \mathbb{R}P^1$).*

Proof. 1. $\Rightarrow$ 2. If γ is not totally real, there would exist an open set $U \subset \mathbb{R}P^1$ such that $\mathrm{card}(\gamma_{\mathbb{R}}^{-1}(x)) < 2vol(\Delta)\ \forall x \in U$; then we would have also $\mathrm{card}(g^{-1}(x)) < 2vol(\Delta)\ \forall x \in U$, which implies that $\mathrm{vol}(\mathrm{im}(g)) < 2\pi vol(\Delta)$.
2. $\Rightarrow$ 1. Since the map γ is totally real, we have that the fibers of g are generically of cardinal $2vol(\Delta)$, which implies that $\int_{\mathcal{A}_{\mathbb{R}A}} |k| = \mathrm{vol}(\mathrm{im}(g)) = 2\pi\mathrm{vol}(\Delta)$, i.e., the real amoeba $\mathcal{A}_{\mathbb{R}A}$ has maximal curvature. □

Remark 2.1. Notice that if $\mathcal{A}_{\mathbb{R}A}$ has maximal curvature, then:

- For $x \in \mathbb{R}P^1$, $\gamma^{-1}(x) \subset \mathbb{R}A$, which implies $F \subset \mathbb{R}A$ since $F = \gamma^{-1}(\mathbb{R}P^1)$ ([5]) and therefore $F = \mathbb{R}A$.

- The only possibilities for $\mathcal{A}_{\mathbb{R}A}$ to have inflection points are at "pinching points" (cf. [5]), which are also the only possible singular points of $\mathcal{A}_{\mathbb{R}A}$ (we will see that in fact if $\mathcal{A}_{\mathbb{R}A}$ has maximal curvature, then $\mathcal{A}_{\mathbb{R}A}$ is smooth).

3. Simple Harnack curves

Definition 3.1. 1) A real algebraic smooth curve $A \subset (\mathbb{C}^*)^2$ with Newton polygon Δ is called maximal (or an M-curve) if the number of connected components of $\mathbb{R}\bar{A}$ is $g+1$, where g is the genus of $\bar{A}$ (or the number of integer points in the interior of Δ).
2) A is said to be "in weak maximal position" if $\bar{A}$ is smooth and transverse to T^∞_Δ and if for $1 \leq i \leq n$, $\bar{A} \cap \mathbb{R}l_i = \bar{A} \cap l_i$ (which means that $\bar{A} \cap T^\infty_\Delta$ is totally real).

Let us look at the following properties for an M-curve A in weak maximal position.

(1) There exists a connected component $C \subset \mathbb{R}\bar{A}$ such that for $1 \leq i \leq n$, we have $C \cap \mathbb{R}l_i = \bar{A} \cap l_i$ (i.e. $C \cap \mathbb{R}l_i$ is made of d_i points).
(2) Condition (1) is verified and for $1 \leq i \leq n$, $C \cap \mathbb{R}l_i$ is contained in an arc C_i of C, such that $C_i \cap C_j = \emptyset$ for $i \neq j$.
(3) Condition (2) is verified and there is an orientation of $\partial\Delta$ such that C is cyclicly in maximal position (see [5]).

Notice that condition (3) is the definition of a "simple Harnack curve", cf. [5].

Remark 3.1. (1) It is proved in [5] that a simple Harnack curve verifies also the following property:
For $1 \leq i \leq n$ the order of the points $C_i \cap \mathbb{R}l_i$ is the same on C_i and $\mathbb{R}l_i$ (for some orientations).
(2) An alternative to conditions (2) and (3) is given by the following lemma:

Lemma 3.1. *With the above notations, assume that A is an M-curve in weak maximal position and that condition* (1) *is fulfilled. Then the following conditions are equivalent:*

(a) *The curve A is a simple Harnack curve (i.e., conditions* (2) *and* (3) *are realised)*
(b) *The amoeba $\mathcal{A}_{\mathbb{R}A}$ is smooth.*

Proof. (a) $\Rightarrow$ (b) is proved in [5] (Corollary 9).
(b) $\Rightarrow$ (a) We have to prove that conditions (2) and (3) above are fulfilled if the amoeba $\mathcal{A}_{\mathbb{R}A}$ is smooth. Assume by absurd that (2) is not fulfilled for a component l_i of T^∞_Δ. Using the moment map, we can see (topologically) the amoeba $\mathcal{A}_{\mathbb{R}A}$ as a subset of $\overset{\circ}{\Delta}$. If $C \subset \mathbb{R}A$, set $\tilde{C}$ for its image in $\overset{\circ}{\Delta}$.
If condition (2) is not fulfilled , there would exist an arc $\tilde{C}_1$ of $\tilde{C}$ joining the side $\tilde{l}_i$ of $\partial\Delta$ to another one $\tilde{l}_j$ and such that if $\overline{\tilde{C}_1} \cap \tilde{l}_i = \{P\}$, there are points of $\tilde{C} \cap \tilde{l}_i$ on both sides of P. Therefore $\Delta \setminus \tilde{C}_1$ has two connected components, and there must exist an arc $\tilde{C}' \subset \tilde{C}$ joining theses two components hence intersecting $\tilde{C}_1$, which implies that the amoeba $\mathcal{A}_{\mathbb{R}A}$ is not smooth.
The proof of condition (3) is similar. $\square$

Proposition 3.2. *If A is a simple Harnack curve, then $\mathcal{A}_{\mathbb{R}A}$ has maximal curvature (Definition 2.1).*

Proof. The logarithmic Gauss map (restricted to the real part) $\gamma_{\mathbb{R}} : \mathbb{R}\bar{A} \to \mathbb{R}P^1$ is totally real (i.e., the cardinality of each fiber is $2vol(\Delta)$) (see Lemma 2.1). Therefore it is also true for the Gauss map $g : \mathcal{A}_{\mathbb{R}A} \to \mathbb{R}P^1$ ($\gamma_{\mathbb{R}} = g \circ L$) since the hypothesis implies that $L|_{\mathbb{R}A}$ is a bijection onto $\mathcal{A}_{\mathbb{R}A}$; then the total curvature of $\mathcal{A}_{\mathbb{R}A}$ is $\text{vol}(\text{Im}(g)) = 2\pi vol(\Delta)$. □

Theorem 3.3. *Let $A \subset (\mathbb{C}^*)^2$ be a smooth real algebraic curve. Assume that $\mathcal{A}_{\mathbb{R}A}$ has maximal curvature. Then A is a simple Harnack curve.*

Proof. Let us first prove the theorem under the supplementary assumption that the real amoeba $\mathcal{A}_{\mathbb{R}A}$ is smooth. Let p be the number of compact connected components of $\mathcal{A}_{\mathbb{R}A}$. Let t be the cardinality of $\mathbb{R}\bar{A} \cap T_{\Delta}^{\infty}$, $s = \sum d_i = \text{card}(\partial\Delta \cap \mathbb{Z}^2)$. Set $g = \text{card}(\overset{\circ}{\Delta} \cap \mathbb{Z}^2)$; g is the genus of the curve $\bar{A}$.

i) By Remarks 2.1 and the smoothness assumption, the real amoeba has no inflection points. Let us prove that since $\mathcal{A}_{\mathbb{R}A}$ is smooth and has no inflection points, one has

$$\int_{\mathcal{A}_{\mathbb{R}A}} |k| \leq 2\pi p + \pi t \tag{2}$$

There are exactly t arcs of $\mathcal{A}_{\mathbb{R}A}$ which are not compact, each one of total curvature $\leq \pi$ (since it has no inflection points), which proves (2) since each compact component have total curvature equal to 2π.

ii) Recall Pick's formula :

$$vol(\Delta) = g + s/2 - 1 \tag{3}$$

The hypothesis of maximal curvature and Pick's formula imply:

$$\int_{\mathcal{A}_{\mathbb{R}A}} |k| = 2\pi vol(\Delta)) = 2\pi g + \pi s - 2\pi$$

which gives, applying (2):

$$2\pi g + \pi s - 2\pi \leq 2\pi p + \pi t. \tag{4}$$

Let us prove that A is an M-curve in weak maximal position. We have necesseraly $t > 0$ because otherwise $s \geq 6$, because the integer length of each side of Δ would then be even, which is not possible by (4) since $p \leq g + 1$. We have therefore $p \leq g$.

(a) Let us first prove that $t = s$ ($t \leq s$ by definition). If we assume by absurd that $t \leq s - 1$, we get from (4) that $2\pi g \leq 2\pi p + \pi$, or $2g \leq 2p + 1$ which implies $g = p$. As above, using the moment map, we can see topologically the amoeba $\mathcal{A}_{\mathbb{R}A}$ as a subset of $\overset{\circ}{\Delta}$. Let us denote by c_{ij} ($i \neq j$) an arc of $\mathcal{A}_{\mathbb{R}A}$ (if it exists) joining $\tilde{l}_i$ to $\tilde{l}_j$; if α_{ij} denotes the angle between $\tilde{l}_i$ and $\tilde{l}_j$, then the total curvature of c_{ij} is α_{ij}. Since $g = p$, there exists only one connected component $C \subset \mathbb{R}\bar{A}$ such that $C \cap T_{\Delta}^{\infty} \neq \emptyset$. Therefore if C intersects

k components l_r ($k \leq n$) of the divisor T^∞_Δ, there exist k arcs $c_{rr'}$ of $\mathcal{A}_{\mathbb{R}A}$ such that each side $\tilde{l}_r$ intersects two of them. Since the sum of the angles between two consecutive sides of Δ is $(n-2)\pi$, we have $\sum \alpha_{ij} \leq (n-2)\pi$. Now there is at most $t-n$ other non compact arcs in $\mathcal{A}_{\mathbb{R}A}$, each one of total curvature $\leq \pi$ (since they have no inflection points), which gives

$$2\pi g + \pi s - 2\pi = \int_{\mathcal{A}_{\mathbb{R}A}} |k| \leq 2\pi g + \pi(t-n) + \pi(n-2) = 2\pi g + \pi t - 2\pi$$

or $s = t$, contrary to the hypothesis.

(b) The relation (4) implies now that $2\pi g + \pi s - 2\pi \leq 2\pi p + \pi s$ (since $t = s$) which implies $g - 1 \leq p \leq g$. But since $p \geq g - 1$, there is at least one arc c of $\mathbb{R}\bar{A}$ which intersects two different components l_i and l_j of T^∞_Δ; therefore the total curvature of the corresponding arc of $\mathcal{A}_{\mathbb{R}A}$ is $< \pi$. We have then $2\pi g + \pi s - 2\pi < 2\pi p + \pi s$, which implies that necessarily $p = g$.
We have then that A is an M-curve in weak maximal position satisfying conditon (1). Now A is a simple Harnack curve by Lemma 3.3, since we have assumed $\mathcal{A}_{\mathbb{R}A}$ is smooth. □

4. On the smoothness of the real amoeba

To achieve the proof of Theorem 3.5, We must prove that under the hypothesis of maximal curvature, the real amoeba $\mathcal{A}_{\mathbb{R}A}$ is smooth. Since A is smooth and that $\mathcal{A}_{\mathbb{R}A}$ does not have smooth inflection points (Remark 2.1), the only *a priori* possible inflection points (or singular points) for $\mathcal{A}_{\mathbb{R}A}$ are pinching points (see [5]). At a pinching point P there are two smooth branches of $\mathcal{A}_{\mathbb{R}A}$ crossing at P, each one with an inflection point at P (see [5]). Let α ($0 \leq \alpha \leq \pi/2$) be the angle between these two branches at P.

Lemma 4.1. *Assume that $\mathcal{A}_{\mathbb{R}A}$ has a pinching point P. Then the real amoeba $\mathcal{A}_{\mathbb{R}A}$ does not have maximal curvature, i.e., there exists $\epsilon > 0$ such that*

$$\int_{\mathcal{A}_{\mathbb{R}A}} |k| \leq 2\pi g + \pi(s-2) - \epsilon.$$

Proof. Let $B_P \subset \mathbb{R}^2$ be a ball centered at P small enough such that it contains no other singular point of $\mathcal{A}_{\mathbb{R}A}$ than the point P. Since P is a pinching point and the curve A is smooth, we have that $L^{-1}(P) = \{P_1, P_2\}$ where $P_i \in \mathbb{R}A$ are two points in different quadrants of $(\mathbb{R}^*)^2$. Then $L^{-1}(\mathcal{A}_{\mathbb{R}A} \cap B_P)$ has two connected components C_1 and C_2 (in different quadrants of $(\mathbb{R}^*)^2$) such that $P_1 \in C_1$ and $P_2 \in C_2$. Let γ_1 be the logarithmic Gauss map restricted to C_1. Since C_1 has by hypothesis a logarithmic inflection point at P_1, the map γ_1 has a local extremum at P_1 (in a chart of $\mathbb{R}P^1$). If $\alpha_1 = \gamma_1(P_1)$, where $\alpha_1 \in]-\pi/2, \pi/2[$, and if α_1 is for instance a local maximum at P_1 for γ_1, then there exists $\varepsilon > 0$ such that for $\theta \in]\alpha_1, \alpha_1 + \varepsilon[$, $\gamma_1^{-1}(\theta)$ has two non real (conjugate) points in $L^{-1}(B_P)$.
Then the logarithmic Gauss map $\gamma : A \to \mathbb{C}P^1$ is not totally real, and the amoeba $\mathcal{A}_{\mathbb{R}A}$ does not have maximal curvature (Lemma 2.1). □

Remark 4.1. If the real amoeba $\mathcal{A}_{\mathbb{R}A}$ has not maximal curvature, it may have a pinching point P. However, if α is the angle between the two branches of $\mathcal{A}_{\mathbb{R}A}$ crossing at P, it is possible to prove that the angle α is necessarily > 0.

References

[1] I. Gelfand, M. Kapranov and A. Zelevinski, *Discriminants, resultants, and multidimensional determinants*, Mathematics: Theory & Applications, Birkhäuser, 1994.
[2] M. J. Greenberg, *Lectures on algebraic topology*, Benjamin, 1967.
[3] A. G. Hovanskiĭ, Newton polyhedra and the genus of complete intersections, *Funktsional. Anal. i Prilozhen.* **12** (1978), no. 1, 38–46.
[4] R. Langevin, Courbure et singularités complexes, *Comment. Math. Helv.* **54** (1979), 6–16.
[5] G. Mikhalkin, Real algebraic curves, moment map and amoebas, *Annals of Math.* **151** (2000), 309–326.
[6] G. Mikhalkin, Amoebas of algebraic varieties, arXiv preprint, math.AG/0108225
[7] G. Mikhalkin and H. Rullgård, Amoebas of maximal area, *Intern. Math. Res. Notices*, **9** (2001), 441–451.
[8] M. Passare and H. Rullgård, Amoebas, Monge-Ampère measures and triangulations of the Newton polytope, *Duke Math. J.* **121** (2004), 481–507.
[9] J.-J. Risler, On the curvature of the real Milnor fiber, *Bull. London Math. Soc.* **35** (2003), no. 4, 445–454.

Department of Mathematics, Stockholm University, SE-106 91 Stockholm
E-mail address: passare@math.su.se

IMJ, UPMC, 4, place Jussieu, case 247 75252 Paris Cedex 05
E-mail address: risler@math.jussieu.fr

Proceedings of 17th Gökova
Geometry-Topology Conference
pp. 135 – 144

Nash homotopy spheres are standard

Selman Akbulut

Abstract. We prove that the infinite family of homotopy 4-spheres constructed by Daniel Nash are all diffeomorphic to S^4.

0. Introduction

In [D], D. Nash constructed an infinite family of smooth homotopy 4-spheres $\Sigma_{p,q,r,s}$ indexed by $(p,q,r,s) \in \mathbb{Z}^4$, and conjectured that they are possibly not diffeomorphic to S^4. Here we prove that they are all diffeomorphic to S^4. In spirit, Nash's construction is an easier version of the construction of the Akhmedov-Park in [AP], namely one starts with a standard manifold $X^4 = X_1 \cup_\partial X_2$ which is a union of two basic pieces along their boundaries, then does the "log transform" operations to some imbedded tori in both sides with the hope of getting an exotic copy of a known manifold M^4. In Nash's case X is the double of $T_0^2 \times T_0^2$ and M is S^4, in Akhmedov-Park case X is the "Cacime surface" of [CCM] (see [A1] for decomposition of X) and M is $S^2 \times S^2$.

1. Log transform operation

First we need to recall the *log transform* operation. Let X be a smooth 4-manifold which contains a torus T^2 with the trivial normal bundle $\nu(T^2) \approx T^2 \times B^2$. Let φ_p $(p \geq 0)$ be the self-diffeomorphism of T^3 induced by the automorphism

$$\begin{pmatrix} 1 & 0 & 0 \\ 0 & p & -1 \\ 0 & 1 & 0 \end{pmatrix}$$

of $H_1(S^1;\mathbf{Z}) \oplus H_1(S^1;\mathbf{Z}) \oplus H_1(S^1;\mathbf{Z})$ with the obvious basis (a,b,c). The operation of removing $\nu(T)$ from X and then regluing $T^2 \times B^2$ via $\varphi_p : S^1 \times T^2 \to \partial\nu(T)$ is called the p log-transform of X along T^2. In short we will refer this as $(a \times b, b, p)$ log transform. Figure 1 describes this as a handlebody operation (cf. [AY] and [GS]).

2. Handlebody descriptions of $T^2 \times T^2$ and $T_0^2 \times T_0^2$

We follow the recipe of [A1] for drawing surface bundles over surfaces. Here we have the simpler case where the base and the fibers are tori. Figure 2 describes the handlebody of T^2 and its thickening $T^2 \times [0,1]$. Figure 3 is a handlebody of $T^2 \times S^1$. Note that

---

1991 *Mathematics Subject Classification.* 58D27, 58A05, 57R65.
The author is partially supported by NSF grant DMS 0905917.

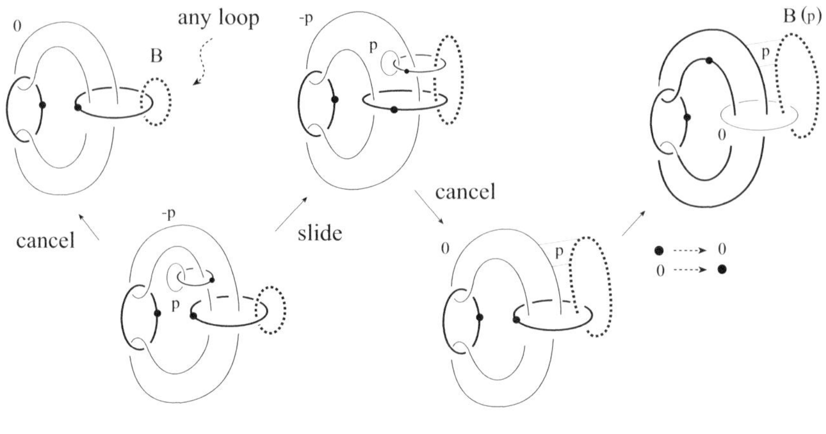

FIGURE 1. p log-transform operation

without the handle v (the green curve) it gives $T_0^2 \times S^1$. So Figure 4 is the handlebody of $T^2 \times T^2$. Note that without 2-handles u and v (the green and blue curves) Figure 4 it is just $T_0^2 \times T_0^2$. By gradually converting the 1-handles of Figure 4 from "pair of balls" notation to the "circle-with-dot" notation of [A2] we get the handlebody pictures of $T^2 \times T^2$ and $T_0^2 \times T_0^2$ in Figure 5 (the first and the second pictures). Then by an isotopy (indicated by the arrow) we obtained the first picture of Figure 6 which is $T_0^2 \times T_0^2$.

3. Nash spheres are standard

Let $X_{p,q}$ be the manifold obtained from $T_0^2 \times T_0^2$ by the log-transformations $(a \times c, a, p)$ and $(b \times c, b, q)$, where a, b, c, d are the circle factors of $T_0^2 \times T_0^2$ indicated of Figure 6. Than Nash homotopy spheres are defined to be:

$$\Sigma_{p,q,r,s} = X_{p,q} \cup_{\phi} -X_{r,s}$$

where ϕ is the involution on $\partial(T_0^2 \times T_0^2)$ flipping $T_0^2 \times S^1$ and $S^1 \times T_0^2$. Notice that if $X^{r,s}$ is the manifold obtained from $T_0^2 \times T_0^2$ by the log transformations $(c \times a, c, r)$ and $(d \times a, d, s)$ then we can identify:

$$\Sigma_{p,q,r,s} = -X_{p,q} \cup_{id} X^{r,s}$$

Theorem 1. $\Sigma_{p,q,r,s} = S^4$

Proof. By using the description of the log-transform in Figure 1, we see that the second picture of Figure 6 is $X_{p,q}$. Now we will turn $X_{p,q}$ upside down and glue it to $X^{r,s}$ along its boundary. For this we take the image of the dual 2-handle curves of $X_{p.q}$ (indicated by the dotted circles in Figure 6) by the diffeomorphism $\partial(X_{p,q}) \xrightarrow{\approx} \partial(X^{r,s})$, and then attach 2-handles to $X^{r,s}$ along the image of these curves. By reversing the log-transform process of Figure 6 we obtain the first picture of Figure 7, which is just $T_0^2 \times T_0^2$, with the dual handle curves indicated! By an isotopy we obtain the second picture of Figure 7, also describing $T_0^2 \times T_0^2$. Now again by using the recipe of Figure 1 we perform the $(c \times a, c, r)$ and $(d \times a, d, s)$ log-transforms and obtain the first picture of Figure 8, which is $X^{r,s}$, with the dual 2-handle curves of $X_{p,q}$ clearly visible in the picture. Now by attaching 2-handles to top of $X^{r,s}$ along these curves we obtain $\Sigma_{p,q,r,s}$, which is also described by the first picture of Figure 8. Here we should mention our convention: when a framing is not indicated in figures it means the zero framing. Now by the obvious handle slides and cancellations in Figure 8 $\rightsquigarrow$ Figure 9 we obtain $\#^4(S^2 \times B^2)$, and the four 3-handles cancel these to give S^4 $\square$.

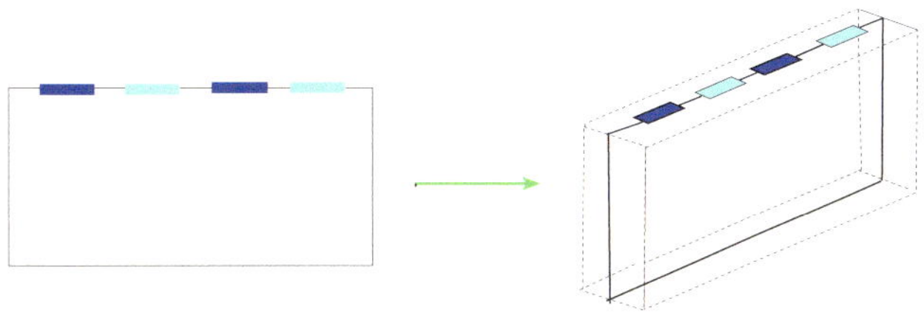

FIGURE 2. $T^2 \to T^2 \times [0,1]$

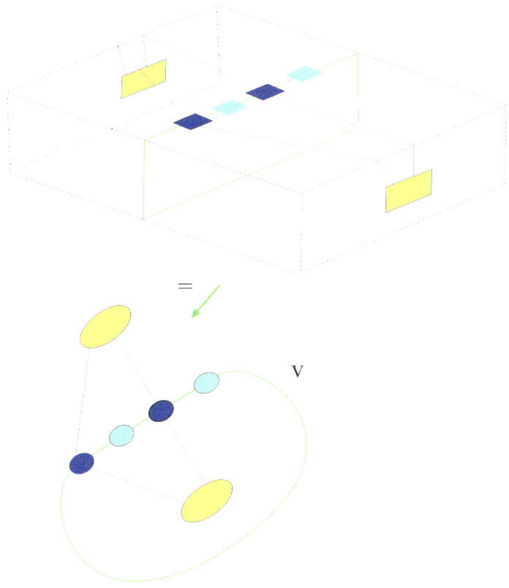

FIGURE 3. $T^2 \times S^1$

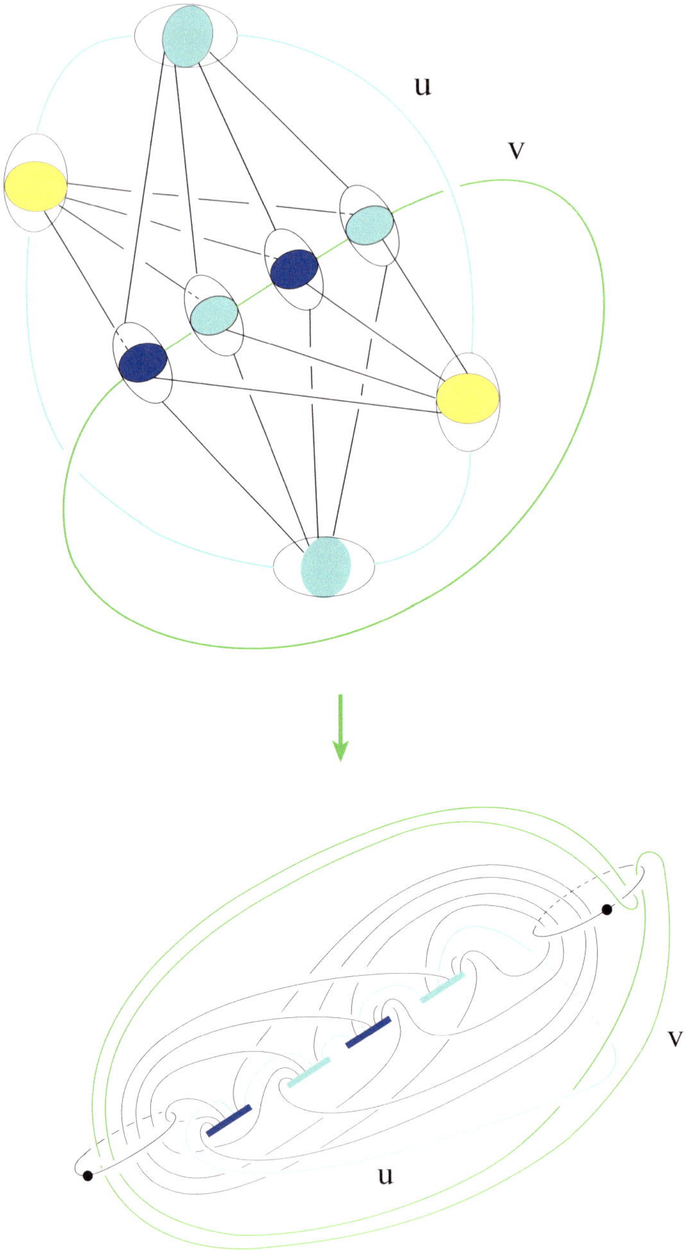

FIGURE 4. $T^2 \times T^2$

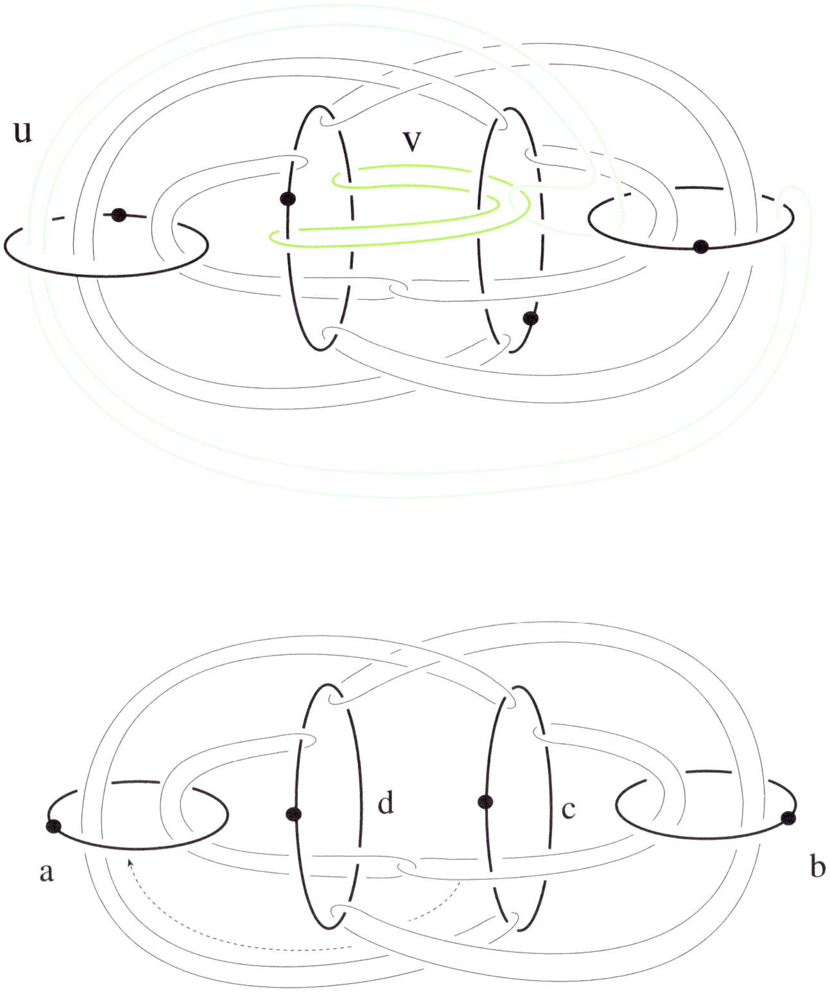

FIGURE 5. $T^2 \times T^2$ and $T_0^2 \times T_0^2$

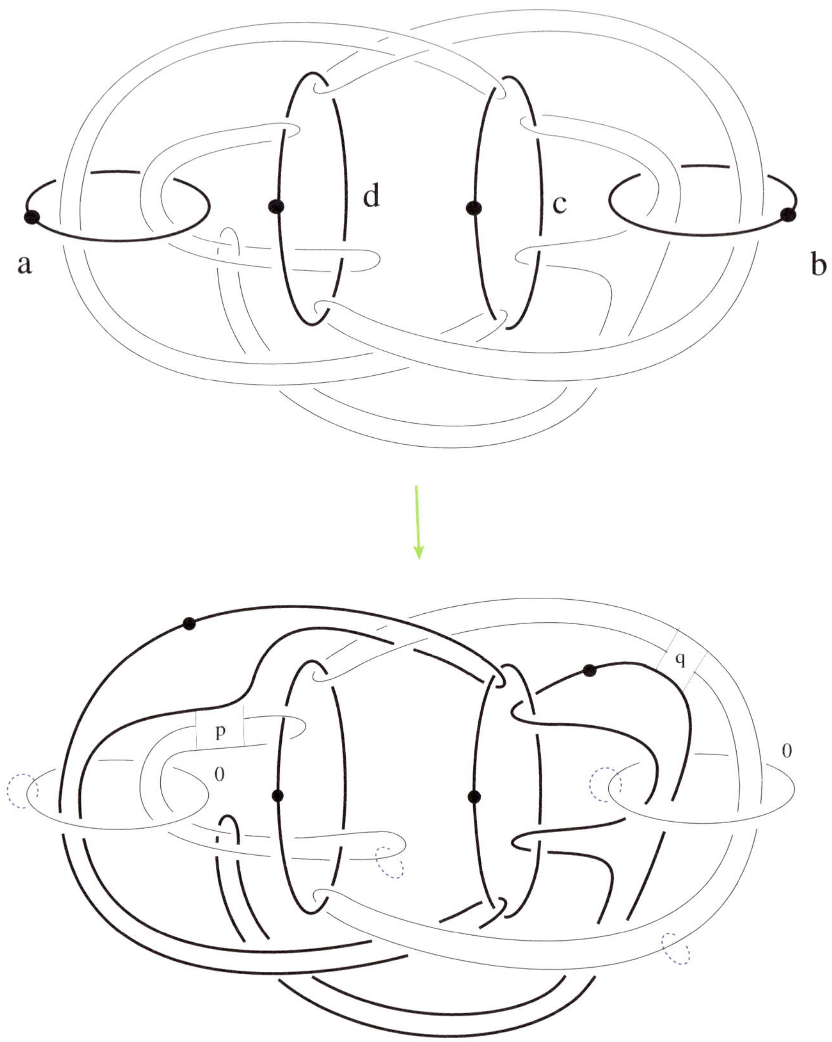

FIGURE 6. log-transformation $T_0^2 \times T_0^2 \to X_{p,q}$

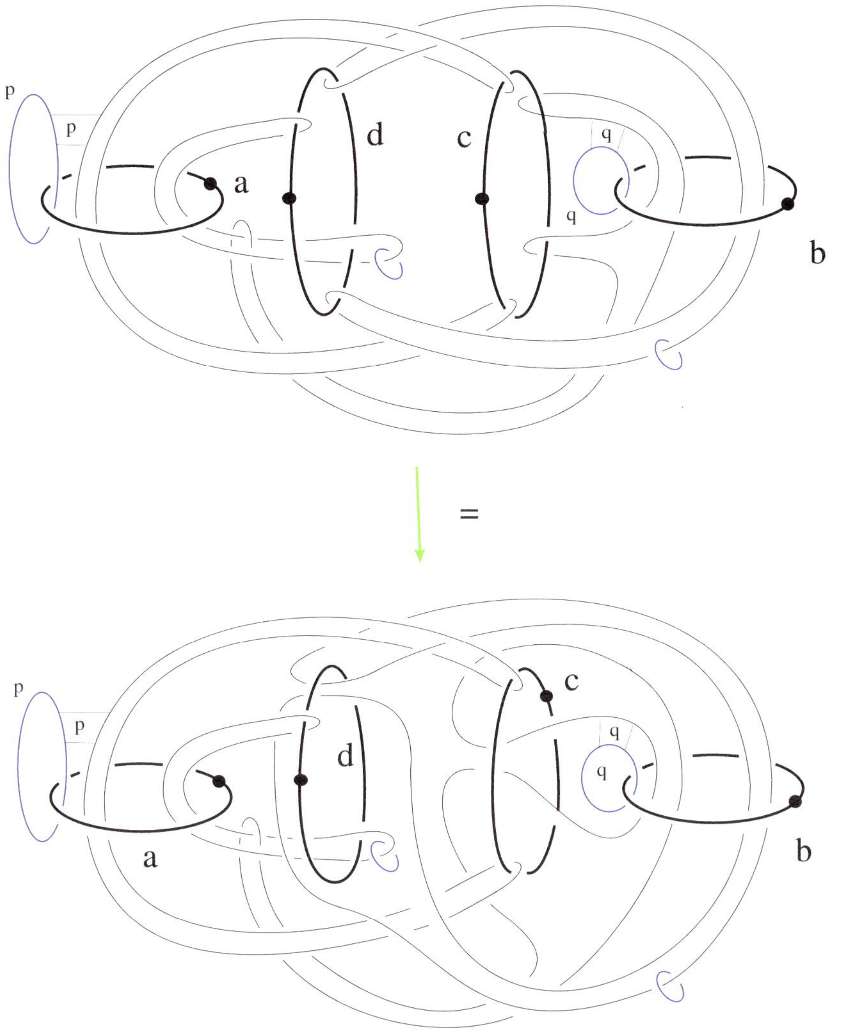

FIGURE 7. log-transformation undone

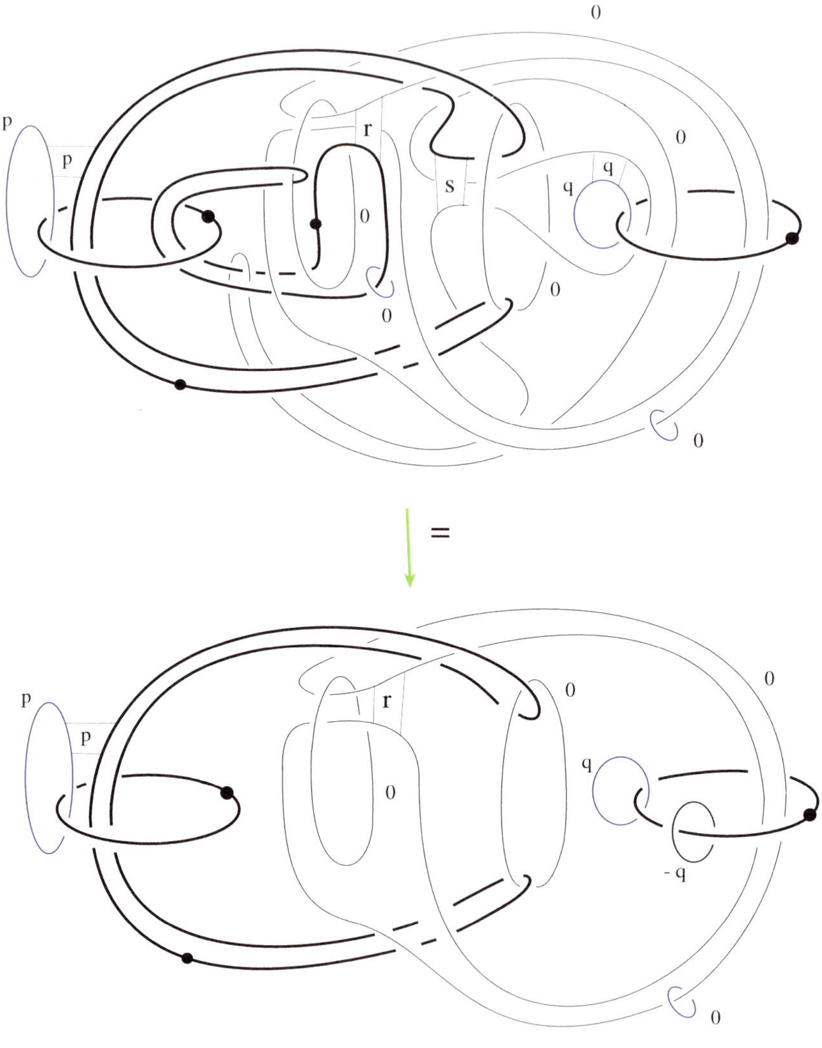

FIGURE 8. $\Sigma_{p,q,r,s}$

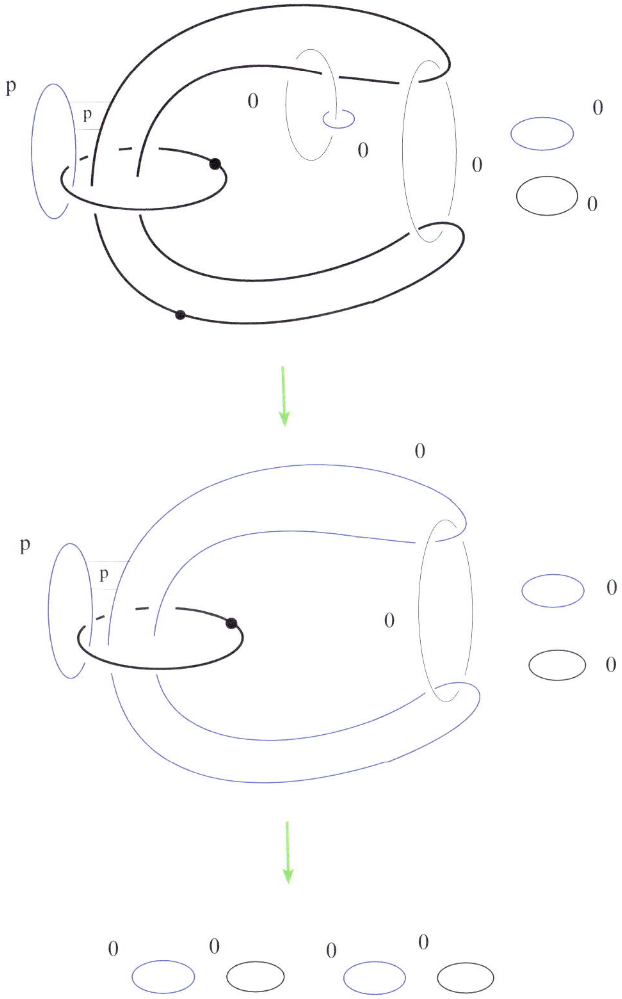

FIGURE 9

References

[A1] S. Akbulut, The Catanese-Ciliberto-Mendes Lopes surface, arXiv preprint, arXiv:1101.3036v1

[A2] S. Akbulut, On 2-dimensional homology classes of 4-manifolds, *Math. Proc. Camb. Phil. Soc.* **82** (1977), 99–106.

[AY] S. Akbulut and K. Yasui, Corks, Plugs and exotic structures, *Journal of Gökova Geometry Topology*, **2** (2008), 40–82.

[AP] A. Akhmedov and B. D. Park, Exotic Smooth Structures on $S^2 \times S^2$, arXiv preprint, arXiv:1006.0293v2

[CCM] F. Catanese, C. Ciliberto and M. Mendes Lopes, On the classification of irregular surfaces of general type with non birational bicanonical map, *Trans. Amer. Math. Soc.* **350** (1998), 275–308.

[D] D. Nash, New homotopy 4-spheres, arXiv preprint, arXiv:1101.2981v1

[GS] R. E. Gompf and A. I. Stipsicz, *4-manifolds and Kirby calculus*, Graduate Studies in Mathematics, 20, American Mathematical Society, 1999.

Department of Mathematics, Michigan State University, MI, 48824
E-mail address: akbulut@math.msu.edu

VOLKAN

17th GÖKOVA GEOMETRY-TOPOLOGY CONFERENCE
May 31 – June 4, 2010

List of Participants

Acu, Bahar	Middle East Technical University, Turkey
Adalı, Seçkin	Middle East Technical University, Turkey
Akbulut, Selman	Michigan State University, USA
Akhmedov, Anar	University of Minnesota, USA
Akyar Moeller, Bedia	Dokuz Eylül University, Turkey
Andersen, Joergen Ellegaard	Aarhus University, Denmark
Argüz, N. Hülya	Middle East Technical University, Turkey
Arslan, Aykut	Michigan State University, USA
Atalan Ozan, Ferihe	Atılım University, Turkey
Auroux, Denis	UC Berkeley, USA
Batakçı, Leyla	Elizabethtown College, USA
Bauer, Stefan	Universität Bielefeld, Deutschland
Baykur, İnanç	Brandeis University, USA
Beyaz, Ahmet	Middle East Technical University, Turkey
Bhupal, Mohan	Middle East Technical University, Turkey
Bourgeois, Frederic	University Libre de Bruxelles, Belgium
Brugallé, Erwan	Institut Mathematiques de Jussieu, France
Ceyhan, Özgür	KdV Institute, University of Amsterdam, Holland
Chen, Weimin	University of Massachusetts-Amherst, USA
Coşkunüzer, Barış	Koç University, Turkey
Çelik, Sinem	Boğaziçi University, Turkey
Çelik, Münevver	Middle East Technical University, Turkey
Dalyan, Elif	Hitit University, Turkey
Degtyarev, Alexander	Bilkent University, Turkey
Demir, A. Sait	Istanbul Technical University, Turkey
Eden, Sinan	Boğaziçi University, Turkey
Efe, Barış	Michigan State University, USA
Erdoğan Demir, Sultan	Bilkent University, Turkey
Etgü, Tolga	Koç University, Turkey
Finashin, Sergey	Middle East Technical University, Turkey

Fock, Vladimir	IRMA, Université de Strasbourg, France
Fomin, Sergey	University of Michigan, USA
Gabai, David	Princeton University, USA
Genç, Özhan	Middle East Technical University, Turkey
Ghiggini, Paolo	Université de Nantes, France
Gören, Yusuf	Boğaziçi University, Turkey
Güçlükan, Aslı	Bilkent University, Turkey
Güğümcü, Neslihan	Middle East Technical University, Turkey
Gürbüzer, Kaan	Dokuz Eylül University, Turkey
İnce, Özgür	University of Rochester, USA
İşim Efe, Melike	Ege University, Turkey
Itenberg, Ilia	IRMA, Université de Strasbourg, France
Kalafat, Mustafa	University of Wisconsin, USA
Karakurt, Çağrı	Michigan State University, USA
Kişisel, Özgür	METU North Cyprus Campus, North Cyprus
Koca, Caner	Stony Brook University, USA
Koçak, Şahin	Anadolu University, Turkey
Korkmaz, Mustafa	Middle East Technical University, Turkey
Lekili, Yankı	University of Cambridge, UK
Mathews, Daniel	Boston College, USA
Maydanskiy, Maksim	Cambridge University, UK
Medetoğulları, Elif	Atılım University, Turkey
Mikhalkin, Grigory	Université de Genève, Switzerland
Oka, Mutsuo	Tokyo University of Science, Japan
Ozan, Yıldıray	Middle East Technical University, Turkey
Önder, Turgut	Middle East Technical University, Turkey
Özer, Arda Buğra	Michigan State University, USA
Özgüner Akyol, Ayşegül	Bilkent University, Turkey
Öztürk, Ali	Abant İzzet Baysal University, Turkey
Pamuk, Mehmetcik	Middle East Technical University, Turkey
Pamuk, Semra	Middle East Technical University, Turkey
Petrunin, Anton	Pennsylvania State University, USA
Ravenel, Douglas	University of Rochester, USA
Risler, Jean-Jacques	IMJ, Université Pierre et Marie Curie, France

Rozhkovskaya, Natasha	Kansas State University, USA
Sağlam, K. Nur	Middle East Technical University, Turkey
Salur, Sema	University of Rochester, USA
Sandon, Sheila	Université de Nantes, France
Shapiro, Michael	Michigan State University, USA
Taşkent, Selin	University of Rochester, USA
Todd, Albert	University of Rochester, USA
Topkara, Mustafa	Mimar Sinan GS University, Turkey
Uyanık, Çağlar	Middle East Technical University, Turkey
Van Horn-Morris, Jeremy	American Institute of Mathematics, USA
Walter, Benjamin	METU North Cyprus Campus, North Cyprus
Williams, Lauren	UC Berkeley, USA
Yalçınkaya, Eyüp	Middle East Technical University, Turkey
Yeniaras, Esra	University of Wisconsin, USA
Zabun, R. Arzu	Middle East Technical University, Turkey
Zarev, Rumen	Columbia University, USA
Zhang, Ping	Eastern Mediterranian University, North Cyprus
Zharkov, Ilia	Kansas State University, USA

Made in the USA
Las Vegas, NV
03 November 2026

57423616R00095